T0073420

Copernicus Books

Sparking Curiosity and Explaining the World

Drawing inspiration from their Renaissance namesake, Copernicus books revolve around scientific curiosity and discovery. Authored by experts from around the world, our books strive to break down barriers and make scientific knowledge more accessible to the public, tackling modern concepts and technologies in a nontechnical and engaging way. Copernicus books are always written with the lay reader in mind, offering introductory forays into different fields to show how the world of science is transforming our daily lives. From astronomy to medicine, business to biology, you will find herein an enriching collection of literature that answers your questions and inspires you to ask even more.

George Vekinis

Physics in the Kitchen

George Vekinis
Athens, Greece

ISSN 2731-8982 ISSN 2731-8990 (electronic)
Copernicus Books
ISBN 978-3-031-34406-0 ISBN 978-3-031-34407-7 (eBook)
https://doi.org/10.1007/978-3-031-34407-7

Image credits: Image at start of Chapter 2 by Marc Manhart from Pixabay; Images at start of Chapters 1, 3–5 by Elf-Moondance from Pixabay; All other images licensed under Creative Commons 3.0

This Springer imprint is published by the registered company Springer Nature Switzerland AG
The registered company address is: Gewerbestrasse 11, 6330 Cham, Switzerland

Contents

About the Author

George Vekinis is a research director and the former head of the Education Office at the National Centre for Scientific Research "Demokritos" in Athens, Greece, and a university lecturer on advanced materials and technological entrepreneurship. He earned a Ph.D. in Physics at the University of the Witwatersrand in South Africa and an M.B.A. (Core) at the Open University, UK. In the past he worked at universities in South Africa and the UK (University of Cambridge) and has served on numerous scientific and technical committees. He has traveled extensively, supervised the research work of over 120 students, and published and lectured extensively on physics, space exploration, materials science and engineering as well as technology transfer and entrepreneurship. His work has been funded by the European Commission, the European Space Agency as well as various industrial entities, and he is the author of two books on entrepreneurship and technology commercialization. He is happily married in Athens with two grown children and two three-legged rescue cats. At weekends, he can reliably be found cooking up a storm in the kitchen.

About the Author

1

Introduction

"Explore the world. Nearly everything is really interesting if you go into it deeply enough."

Richard P. Feynman

G. Vekinis, *Physics in the Kitchen*, Copernicus Books,
https://doi.org/10.1007/978-3-031-34407-7_1

A Physicist Cook's Welcome

The laws of physics are everywhere—from the way we think and breathe and walk to the way our devices operate, and from the way our food is cooked to the way we eat it and get sustenance from it. They are so ubiquitous and all pervasive that they are invisible. We have grown up so immersed in them that we don't even consider them or think about them as we go about our business under their total control. They are hidden in plain sight. Nearly everything happens in nature—the kitchen included of course—according to a plethora of unchanging rules (that's why we call them "laws"), discovered and clarified over many years by thousands of patient and persevering scientists. They are the rules according to which the Earth moves, plants grow, and we stay alive. Ok, we don't yet completely understand everything, but give us time… every day thousands of scientists are working towards a better and better understanding of nature at all levels. From its workings at the atomic and sub-atomic levels right up to the Universe as a whole and its origin. We think up theories and hypotheses and then carry out systematic experiments and careful observations to confirm them or discard them. And we are guided by them in our everyday lives, mostly without realising it.

That's what this book is all about. I'm a research physicist and at the same time I love cooking and experimenting with new combinations of foods, aiming to make interesting meals and cakes in the kitchen. But most of all I love observing the physical phenomena that occur while I'm doing the cooking and seeing how they affect the results. While I'm at it, I make mental notes about how things proceed and may make adjustments or try new ideas depending on what the physics—and the results of my experiments—tell me. It's the scientific method in the kitchen. I don't always succeed in making a gourmet supper, but the physics I see and the experiments I try are definitely exciting! And sometimes it happens that the details of certain phenomena we see in the kitchen are not what we expect or have come to believe.

I do the same in my scientific research. I observe what happens during and after making specific adjustments in my experiments, and the feedback I get guides my next steps. At the same time, I keep an open mind for corrections or even additions to my knowledge. Whereas most of the basic laws of physics are well understood and have been confirmed thousands of times, some lesser phenomena are still under investigation and their meaning is still being debated. And if we find that an existing theory is not completely correct, we are the first to correct it or even discard it. That's how scientific study works. By checking again and again, by going to sleep with questions and waking up with new ideas to test in order to answer them. I often do the same with cooking. If something doesn't work as I hope, I adjust it or change it.

In fact, if I may digress a bit, that's how the two greatest discoveries of modern physics were made. Struggling to account for certain "annoying" experimental and observational discrepancies that couldn't be explained with the knowledge of that time, the great scientists of the beginning of the twentieth century discovered the two greatest fields of modern physics: quantum mechanics[1] and the modern theory of gravitation (the general theory of relativity).[2] The first gave us a clearer picture of the atomic world, and it also gave us electronics, computers, and a myriad other technological marvels that we can now hardly imagine living without (including in the kitchen), while the second gave us a much better understanding of the birth, development and dynamics of the Universe and all the stars and other bodies within it.

But enough digression. Let's go back to the kitchen and the physics that abounds in it. All the phenomena that we observe while cooking obey the laws of physics, as does everything that happens around the kitchen obey the laws of physics. Actually, everything obeys the laws of physics, at least on this Earth, and we believe everywhere else in the universe too. This little book is an attempt to point out and clarify a few of these laws that govern our everyday lives. And what better place to search for and observe them than in our pots and pans and the various devices we keep in our own kitchen. I wrote it because I love observing how ingredients (themselves made by and governed by the laws of physics) heat and mix and combine to make a meal that will in turn (hopefully) excite and please our brain via what else but our physics-obeying taste buds, smell sensors, eyes, and even ears.

The modern kitchen is an exciting place. It's full of devices, gadgets, and machines created for our benefit. It's the nexus of so many of our daily activities that the mind boggles when we start considering them all. It's where energy, water, and raw materials meet and are tamed by our own energy and ingenuity to create food for our pleasure and sustenance. And it's where we exploit and use—usually unknowingly—a whole range of the laws of physics.

Consider something as simple as water coming out of a tap. How does that happen? There is a fundamental law of physics that says that any system if left alone will always try to reduce its internal energy.[3] Because the water

[1] Ludwig Boltzmann, Max Planck, and Albert Einstein to begin with, followed by Niels Bohr, Louis de Broglie, Max Born, Paul Dirac, Werner Heisenberg, Wolfgang Pauli, Erwin Schrödinger, and later, Richard Feynman. And many others too.

[2] Albert Einstein, single-handedly.

[3] In thermodynamics, we'll see later, there is an equivalent law for heat (the "second law") which says that heat can only flow from a place of high to a place of low temperature. If you want to reverse the flow, you have to input energy.

in the pipe is under a high pressure (about 6 times atmospheric pressure and the poor old pipes are straining to contain it, sometimes failing by rupture or leakage), it contains a large amount of energy. As soon as the tap is opened, water gushes out, because this way the energy of the system (the pipe and the water in it) is reduced. The water we see gushing out is pushed out by the water behind it, which is still under pressure in the pipe. Just like a mass of people pushing to go through a narrow gate into a stadium. As soon as you get through the gate, you feel the pressure behind you decreasing and you are free. The same happens in the water pipe. As soon as the water emerges, it is free and its pressure drops to that of the surroundings. Of course, it is immediately acted upon by gravity, which forces it downwards. If there wasn't any gravity, it would just dance around aimlessly, splashing about, happy to be free.

And, guess what. Exactly the same principle governs the movement of electrons through a wire. There is a force pushing them forward (we call it the "potential difference" or voltage) which forces a few electrons at a time[4] out of the end of the wire and into the electric circuit it is connected to, where its charge (negative, by convention) is used as a unit of energy to carry out some work.

Every time I cook, I enjoy seeing how many physical phenomena are involved in preparing a simple meal or just a cup of tea. Even a simple action in the kitchen such as boiling water in a kettle utilizes many physical phenomena and laws of physics. First, operating the switch forces two pieces of metal (copper) to touch by exerting a force on them which closes an electric circuit. This allows electrons to flow (again in order to minimize the energy of the system) through the wires to the kettle, heating the element, which is made of a material that resists electron flow. This resistance to electrons (something like friction—rub your hands together vigorously and see how they heat up) makes the element heat up. The element then passes ("conducts") the heat to the metallic base of the kettle and from there to the water above it. The water molecules begin to vibrate more and more violently till they can't take it any longer and start escaping as steam!

Read the above paragraph again and count how many physics laws are involved just to boil water! Our finger exerts a force on the switch which snaps shut. Both these actions involve forces which distort the materials involved: our finger (via our muscles) and the switch materials. Such mechanical distortions obey, and are described by the physical laws of elasticity. Electrons "jumping across" the touching metals obey the laws of quantum mechanics,

---

[4] Well, hardly a few … maybe a few trillions per second, depending on the application. If it's for heating, it's actually a heck of a lot—see later.

while electrons flowing around an electric circuit do so according to the laws of electricity. When we heat the element and when we heat the water which boils and evaporates, what happens obeys the laws of thermodynamics. Physics is indeed everywhere.

But Wait—Is It Physics or Chemistry in the Kitchen?

When I was writing this book, I (often) had to answer the question "but isn't cooking more to do with chemistry than with physics?" Well, I admit it's a good question and it's a good idea to clarify it at the outset. Chemistry deals with the properties of molecules, especially when they react together to give something different, like a solution or another compound. But physics explains *how* such reactions and interactions take place. In this respect all chemical processes obey and are underpinned by physical laws, and specifically the rules of quantum mechanics. In the case of chemical reactions, chemistry tells us what is taking place, while physics explains why and how it's taking place. To a good approximation chemistry describes the results of interactions between electron "orbitals" which are defined by quantum mechanics. These "orbitals" have nothing to do with orbits, but are just regions where electrons can be found around atoms. Depending on which orbitals they belong to, the electrons have different, distinct energies. Actually, the word "quantum" ("packet" in Latin) was coined because the energy of each orbital is always some multiple of a very small increment of energy.

Before we go on, I think we need a clarification about chemical bonds and I'll try to keep it simple. All organic molecules, including the proteins, sugars, and everything else we use as food, are made up of atoms which are bonded together in some way. There are many types of bonds in organic materials, but the basic ones that concern us here are "covalent bonds" made by "sharing electrons" (or, more correctly, "interactions between electron orbitals"), which can be moderately to very strong, "ionic" bonds which rely on electrostatic attraction between positive and negative ions (atoms) and "hydrogen bonds," which involve hydrogen atoms and are much weaker. Hydrogen bonds are very widespread in organic materials like foods and a special type of them is often found in small volatile aromatic molecules of herbs, coffee, tea, etc. Moderate strength covalent bonds generally occur between smaller satellite molecules and a main strong core molecule in proteins and other similar structures, as well as between the amino acids that make up the long protein molecules. However, amino acids and other basic molecules always have a backbone made up of carbon atoms which are held together by strong covalent bonds. During cooking, as we increase the total energy input (increasing temperature and time), the hydrogen bonds break up ("dissociate") first, followed by the moderate covalent bonds, and only at high temperatures (that we usually avoid in cooking) do the backbone carbon molecules break up.

Very often, as soon as some bonds break up, new ones form. When a "solute" molecule dissolves in a "solvent" liquid, the solute molecule breaks up and forms new bonds with the solvent molecule.

A lot of chemistry takes place during cooking, but cooking is much more than reactions (or interactions) between molecules. While chemical reactions do take place when we mix and heat ingredients, it is the laws of physics (heat distribution, electrical interactions, orbital interactions, diffusion, etc.) that hold sway over everything that happens in the pot, and in the kitchen in general. From the moment you put a pot on the stove to the moment you pour out a drink and swallow your food, you are, often without thinking about it, exploiting or obeying physical laws.

In fact, when cooking, we often try to avoid chemical reactions. While we exploit and use chemical solutions and try to "brown" certain foods, more often than not we try to preserve many of the properties (the aroma, the taste, the structure) of the raw materials we put in, and try to blend them and find a balance between them which will provide culinary pleasure. In some cases, we do encourage (and control) certain chemical reactions during cooking, such as a slight "caramelisation" when stir frying of onions or leeks, which slightly alters their taste, aroma, and structure. Or we use an acid or the protein of an egg to encourage lipids (fats and oils) to stick to water. Furthermore, one of the pillars of a sumptuous meal, a smooth sauce, does involve chemistry which, in actual fact, is the same chemistry that we use to make plastics: polymerisation reactions. We'll discuss these aspects later and delve more deeply into the corresponding mechanisms in order to understand the physical processes that make them happen.

In a nutshell, this is the main objective of this book. To delve into, wonder at, and elucidate many apparently simple phenomena that occur in the kitchen, things that we usually take for granted. Through numerous examples, I'll try to show you some of the physical phenomena and laws that hide behind what happens in the kitchen, and show how they actually arise. Along the way I hope to be able to give you a fresh perspective on one of the most satisfying human inventions and pursuits, cooking. If you are, like me, an eternally inquisitive type of person and an aspiring and curious cook (forever experimenting, sometimes to the consternation of my family), I hope you'll enjoy the never-ending cascade of wonderful insights that we can get on the way to making that lovely casserole.

A Few Words on the Structure of the Book

In an attempt to ensure some logical sequence and a good physics grounding on the basic concepts (which I'll refer to throughout the book), I have started with an introduction to energy and the laws of "thermodynamics", the science of heat. Many readers may already have a good basic understanding of such matters, and I beg your indulgence. In any case, I'll take you on a brief journey of the basics of energy—the basis and currency of everything—and how it is used in the kitchen, mainly in the form of heat and electricity. We'll look at how energy originates and how it gets transformed all the way from the power station until it becomes heat in the kitchen, where some of it is used to cook and some of it is lost to the environment.

Once we have established the basics of heat energy, we'll consider physical aspects of the basic ingredients that go into cooking, starting with the amazing properties of water. We'll look at coffee and tea, the colours of foods, and lots of other unusual aspects of the raw ingredients.

Next, we'll consider the physics and chemistry that goes on in the pot, the frying pan, the oven, and anywhere else where our raw ingredients can be turned into a hopefully palatable meal. This will form the main part of the book and it will include examples of actual cooking which, I hope, will help you see food preparation with a fresh eye and even surprise you in some ways. Here I discuss various cooking tasks, in some detail at times, and describe how stews, sauces, fries, and roasts take on their colour, aroma, and taste. I have attempted to incorporate cooking methods from various parts of the world but, being Greek myself, I have mainly focussed my discussion on Eastern Mediterranean cooking. By the way, I have only used examples and descriptions of foods using natural ingredients and do not discuss any of the numerous artificial agents used for producing ready-made meals, such as artificial emulsifiers, acidification agents, and all the chemistry that goes into making nice-looking and tasty, but hardly natural (or naturally nutritious) food.

Next, we'll look at many of the ubiquitous kitchen appliances, devices, machines, and materials in a modern kitchen. There will be quite a few further surprises there too. By necessity, some of the discussions here may duplicate certain comments I made during the cooking chapters, but they are based on a more technical viewpoint.

Finally, we'll talk about various miscellaneous subjects and odd aspects of cooking and working in the kitchen that don't fit anywhere else, but are nevertheless intriguing and surprising as well.

Nearly all the chapters can be read individually and in a random order, but I think they'll be more satisfactory if you read them approximately in the order presented. If you have a good grounding in physics, then you could skip the first part on energy, thermodynamics, diffusion, etc. However, if you persevere with it, you may also be surprised at certain nuances that are often forgotten or taken for granted.

I should mention that experienced cooks and chefs may find some of my descriptions of cooking processes too simplistic, so I hasten to emphasise that the book is not meant to be a treatise on the methods of cookery, but an attempt to elucidate the underlying physics.

Because many of the physical phenomena I discuss occur again and again in various situations in the kitchen (and especially in cooking), it has been necessary sometimes to repeat certain descriptions, but to avoid tiring the reader, I always try to emphasise the extra dimensions that come into play in each situation, so please bear with me.

Before I proceed, I must ask forgiveness from scientists and purists for the frequent over-simplifications I have used in order to make my explanations of physical phenomena a bit easier and clearer. I have done this in an effort to avoid confusion and jargon. To atone for this, I have included more precise explanations, related aspects, and other interesting bits of information in footnotes and information boxes, where I thought it would enable a deeper understanding and complement the points made in the text. In addition, I have left out nearly all the equations describing the phenomena I discuss, once again to avoid confusion and concentrate only on the physics. In some cases, this means that I only mention and briefly describe the relevant law without going much deeper.

Obviously, I have not attempted to include or discuss in any detail advanced phenomena such as energy perturbations or waves or certain quantum mechanical effects which may occur during cooking and elsewhere in the kitchen, apart from stating them. This is simply because they are too complicated and in any case wouldn't add much to the discussions of the main physical phenomena we encounter when cooking. If anyone wishes to dig deeper, I suggest they consult one of the many good textbooks on physics.

During the preparation and writing of the book I had the pleasure of receiving and discussing many ideas and tips from my wife Gwen and my children Andrew and Stefani, and I thank them for those gems. I especially enjoyed the many heated discussions on what is worth including and how to go about it.

Finally, I want to stress that all the things I have written and discussed in this book are my own opinions and interpretations of the physical phenomena that occur in the kitchen, some of them perhaps not uncontroversial, and I have done so to the best of my understanding and ability based on publicly available knowledge. It goes without saying that everything I have written is my responsibility and mine alone.

So, without further delay, let's enter the world of physics in the kitchen. And the very first item on the menu and the most salient characteristic of all our activities in the kitchen is the fact that everything is driven by energy.

2

Building an Appetite—Energy: The Currency of Life, Food, and ... Everything Else

"Everything is energy"
Albert Einstein

G. Vekinis, *Physics in the Kitchen*, Copernicus Books,
https://doi.org/10.1007/978-3-031-34407-7_2

What is energy? In everyday parlance we understand it as something that gives us the ability to do something: walk, run, heat, dig, breath, cook, blow air, remove air, work on a laptop, etc. It's actually not a bad definition and quite close to the more or less official definition: "Energy is the basic physical entity that enables a system to do useful work". A system could be a machine, a person, or a device of any type. And work means any action whether it has a purpose or not. Usually, we cannot observe energy directly, but we can infer its presence by observing the work that it enables. Nobody and nothing can do any work and nothing can be produced without some energy input that allows it to happen. So, we can say that energy is the currency of life. A limited amount of energy means limited ability to do something and vice versa. Objects which have a lot of energy (either because they are given or produce it) are able to do a lot of work and to carry out more actions. In general, every action everywhere depends on the availability and sharing of energy by the objects that are involved in the action.

Joule: the man and the unit of energy

We measure energy in joules (J) from the nineteenth century English physicist James Prescott Joule who showed that all types of energy (mechanical, electrical, chemical, gravitational, etc.) are basically the same and can be converted from one to another.

A joule is a very small amount of energy, just enough to lift your arm about 10 cm, to light an LED for 1 s, or to lift an apple by one metre. Everyday tasks require thousands or millions of joules of energy. Sleeping for just one minute requires about 6000 J while heating a couple of cups of water in the kettle to boiling temperature uses about 200,000 J! And cooking a meal in a pot might use more than 2 million J!

But what exactly is "work"? There are many ways of defining it, but I like the following: "Work is any process or activity by which something is created or produced or just happens". In the kitchen, energy input enables work to be done so that a kettle can heat water for tea, or you can cut up the ingredients of a salad, or an oven can broil a steak, or a toaster can toast a sandwich. Work enabled by electrical energy may also produce light from a bulb or music from a radio or warmth from a heater.

If energy is the currency of life, work is the tool you get by paying in energy. And in cooking, work is necessary to produce all the intermediate stages as well as the final product. Think about it: when you make a soup or a stew, you pay in energy to prepare ingredients, heat up the hob, and boil water. You also pay in energy to get your mouth to swallow food and to get your stomach and digestive system to work on digesting a meal.

So, energy is used to enable work to produce something. That gives us another angle by which to understand energy and work better. Since, in order to produce something, you have to put some effort into it, the amount of work you need to do is directly related to the amount of effort you need to make. It's also directly related to the spatial extent and the time over which you make that effort. Let's say you need to cut a large piece of hard cheese with a knife. The harder the cheese, the more force you have to exert and the more tired you'll feel because you have expended a lot of energy. But you'll also expend a lot of energy if you have to cut a softer but larger cheese, which will take you longer to cut. In physics, energy is actually measured just like this. If we measure the force and the total distance over which it acts, then the work done and the energy expended are given simply by the product of the force and the distance.

Let's look at another example of work in the kitchen. Making béchamel sauce involves beating (whisking) the butter, flour, and milk mixture over a warm hob (or warm water) for a long time until it thickens. The work you put in is paid for by energy produced in your body and by the hob. The longer you beat the ingredients, the more energy you expend, and the more you can feel the energy being depleted in your muscles, where sugar is slowly converted to lactic acid, giving you that feeling of tiredness. That energy originally comes from metabolism of food, which converts it into usable energy sources like sugar and fat. But only a small proportion of that energy is available for beating the béchamel. As you work, your body still has to expend energy to keep you at the right temperature, to keep your heart and other organs working, to ensure you remain standing and keeping your balance, and to keep your brain thinking about how to move your hand, estimate quantities, prepare the next steps, and so on.

There is no free lunch. We need to provide energy to a system to get it to produce work, but where does this energy come from? In everyday life, and certainly in the kitchen, the energy we use for cooking is available in two forms: electricity and gas for producing a heating flame. Both are used to create heat for cooking, whereas electricity can do many more types of work, such as light the room, beat a cake mixture, squeeze an orange, remove heat from the fridge, cool the room, wash the dishes, or move the air in the oven. Our body is also a system that does work and our energy comes from burning food inside us to produce energy to drive muscles, nerves, and much more.

In this last paragraph alone, I already mentioned a number of everyday tasks based on physical phenomena that deserve further physical explanation: create heat, cook food, operate a blender, beat a mixture, remove heat, cool the room, and many other activities. I aim to explain the physical bases of all these and more in the next chapters.

But first, let's consider what exactly is meant by "heat" in any material.

Hot, Dancing Atoms

A very strange but critical physical aspect of the world is that everything moves. Not just figuratively speaking, but absolutely everything actually moves all the time! Even the most static and calm and docile body or object (e.g., a rock, or a pool of standing water, or a piece of paper, or a banana) is actually in constant motion, deep inside. The atoms and molecules it is made of are always vibrating (even in solids, where they are arranged in more or less regular lattices) or vibrating and zooming around randomly (in liquids and gases). They are continuously interacting with each other and jostling with their surroundings. The atoms that make up your cells and all the fluids inside you are all jumping around and shifting and vibrating constantly. All this is completely invisible to us, but has huge consequences for nature.

More than 25 centuries ago, the ancient philosopher Heraclitus put it all quite nicely: "Ta panta rei", i.e., "everything flows, nothing is constant", even though he knew nothing of atoms and molecules. Nowadays we know that all bodies possess internal energy and are always exchanging or sharing their energy with their surroundings. Energy is always being passed around, between molecules and between atoms, just as billiard balls give energy to each other when knocked about. And nowhere more so than during cooking in the kitchen. Violently vibrating atoms in the hot hob (whether an electric plate or a gas flame) knock on and pass their energy to the pot, which passes it to the other atoms inside it, which knock on and pass their energy to the atoms and molecules in the soup. They in turn pass their energy to the air molecules, and so on. Even the wooden spoon used to stir starts vibrating more vigorously by interacting with the hot soup molecules.

Many-faceted energy

Apart from kinetic energy, there are many other types of internal energy depending on the state a body is in or the position in which it is placed. In all such cases energy is gained by the body (or its atoms) acting against some force. For example, if a body is loaded by some force (compression, tension, etc.), we say it contains strain energy. An example is the taut spring in wind-up watches moving its gears. If some mass is lifted above the ground it now contains gravitational energy, as in water reservoirs (dams) used for electricity production. A battery or cooking gas contains chemical energy and is used to light a bulb or cook a meal. Your body also contains lots of chemical energy used to keep you alive and carry out your chores.

All such types of energy are called "potential energy," as the energy stored internally can potentially be used to carry out some work.

The energy in the atoms' movement is called "kinetic" energy (from the Greek "kinesis" which means motion—lots of words in physics come from Greek—the words "energy", "thermal," and "physics" too) and the sum of all the kinetic energy of all the innumerable atoms in a solid, a liquid, or a gas is called its "internal energy" or "heat content". Notice the interchangeability of the words "energy" and "heat" in this case. So, if we increase the temperature of a soup, what we are actually doing is giving more heat energy to the atoms and molecules in the soup, causing them to vibrate and move around even faster. This faster motion is important because it means that there are more interactions between them, thereby producing a better cooking result. We'll discuss this in more detail later (see box).

We have a simple way to measure the amount of heat energy in a material: we measure the physical change it imparts to a carefully designed instrument and call it "temperature". At low enough temperatures like those encountered in cooking (up to about 250 °C) we used to rely on a glass thermometer. Essentially, we measured the expansion of a liquid (mercury or alcohol) in a thin tube (a capillary) due to the vibrations of its own atoms. The more vigorous the vibrations of its atoms, the more a liquid expands and the higher the temperature that will be indicated. Nowadays, we use electronic thermometers which work very differently, not always more accurately, but with greater precision.[1] We'll talk about some of them later in the book.

[1] These are often confused. "More accurate" means "closer to the true value", while "higher precision" means "measuring greater detail" (even if it's not accurate). For example, an electronic temperature measurement of the boiling point of water at sea level may show 99.872 °C which has higher precision, but it is not accurate since we know that the correct value is 100 °C. It is a mistake to express a result with higher precision than that justified by the experiment and normal measurement errors.

When we say something is "hot," we actually imply "hotter than something else," because its atoms (in the case of solids) or molecules (in the case of liquids and gases) vibrate more violently, due to the fact that they contain more energy. The more violent the vibrations and the faster the molecules move, the hotter the material feels. Increasing the amount of motion of the atoms eventually leads to a change of state of the material. For example, as we add more energy to a solid (by heating it), the stronger vibrations of the atoms make the object expand more and more, and at some point, their motion is so violent that the atomic lattice breaks apart, whereupon the atoms start escaping from the solid and float around. The material gradually becomes a liquid (or directly a gas in some cases) and, if we keep on adding energy, even the tenuous interactions between the liquid atoms or molecules[2] become untenable, so that the liquid transforms into a gas with freely moving atoms or molecules. We know this of course as melting and evaporation, respectively, and we'll see more details later when we discuss boiling.

Employing heat to do work

Thermodynamics has a long history starting with the first steam engines built by Hero of Alexandria one of which was strong enough to operate a heavy temple door. As much as 1600 dark years later, starting with the observations of Robert Boyle, Robert Hooke and Edme Mariotte in the seventeenth century and continuing with the work of Jacques Charles, Sadi Carnot and Emile Clapeyron in the 18th and 19th, the "Ideal Gas Law" explained the relationship between pressure P, volume V and temperature T of a gas (or steam) as $PV = kT$ where k is a constant. Anyone trying to inflate a bicycle tire will attest to the fact that, since V is constant, increasing P increases T. Or, in a steam engine, increasing T will increase P which can then be used to move a huge steam locomotive.

Thermodynamics became fully established later in the nineteenth century when the insights of James Joule, Rudolf Clausius, James Maxwell, Ludwig Boltzmann, and others clarified and explained theoretically the three (later 4) laws and their relation to energy and work. The concept of entropy and the statistical treatment of heat laid the ground for major discoveries and applications.

The nature and effects of the atoms' continual vibration and movement and jostling is the subject of the physics of thermodynamics. In other words,

[2] Most liquids are made up of molecules which continuously dissociate (break up) into constituent atoms or smaller molecules and reform. For example, water is constantly breaking up into a hydrogen atom (a proton) and a hydroxyl molecule OH (both of which are gases if free) and then reforming. But all we see is a liquid.

it is the study of heat and its interactions, and the way it is shared between materials and objects. Let's now dig a bit more deeply into this subject.

Immutable Laws

Heraclitus's maxim is particularly pertinent in the kitchen. From the movement of air molecules forced out by the extractor fan to the exchange of the energy of atoms and molecules in a cooking pot, everything moves, nothing stays put. This fact means that in any situation where two bodies are in good contact (for example, food in a pot on a hot plate or on a fire), the atoms and molecules on each side will bump into each other (soup against pot and pot against plate) and thereby share their (kinetic) energy. If we wait long enough all the atoms and molecules in the three bodies in contact (which includes all types of contact, such as between air and pot or soup and pot) will eventually have the same (average) heat energy and will thus have the same temperature.[3] This fact is a very basic tenet of heat physics and we thus call it the Zeroth law of thermodynamics. The ubiquity of this law will become apparent as we move through the book, and we accept it as given.

But thermodynamics is also ruled by three other immutable laws. The first one you may know already: the total amount of energy in a system which is completely thermally isolated (we call it a "closed system") is always constant. This is called the law of conservation of energy, and we also call it the First law of thermodynamics. It is true for any closed system, anywhere. But it is also true for any system in thermal equilibrium (constant average temperature) even if it interacts with its surroundings. For example, if the average temperature of the whole kitchen stays constant, it means that all energy input must equal all the energy output. The same is true for any sub-system within the kitchen. The fridge would remain at exactly the same low temperature if it weren't for the slow ingress of the surrounding heat, so we have to do work (with the compressor, see later) to extract this input. That's why insulation is so important to slow down this ingress. Since we are cooking and we add energy (gas or electricity), the only way to keep the total energy content of the kitchen constant is to extract the same amount of energy via the extractor fan which helps to keep the overall temperature constant.

[3] Assuming perfect contact between the bodies. In actual fact, while a liquid food is generally in very good contact with the pot, the pot is rarely in perfect contact with the hot plate (hob) underneath, so some heat from the hob will be lost to the surroundings and the pot will thus have a lower temperature than the hob. Gas flames may seem to have better "contact" with a pot, but that's rarely the case.

This always seems like a losing battle when cooking in summertime because so much more energy is contained in the hot surrounding air that it sometimes feels unbearable. That's the reason why we tend to avoid boiled foods in summer, as they add energy to the surroundings and increase humidity, and this only increases heat discomfort. In my kitchen you'll never see a soup on the menu in summer. You see, water holds much more energy than air, so a very high relative humidity[4] in the summer means that the air is almost saturated in water and our body cannot cool down by sweating. The whole planet Earth is feeling this effect now because the greenhouse gases we add to the atmosphere have reduced its ability to lose heat to space, so it is not in thermal equilibrium anymore. Indeed, its average temperature is gradually increasing and causing huge climatic instabilities.

The final state of everything

A fundamental tenet of physics is that everything tends towards greater disorder, eventually. We measure a system's disorder by its entropy, and in a thermally isolated system total entropy can never decrease. Of all types of energy, heat has the highest disorder because heating results in wilder motion of the hot atoms and molecules, so all types of energy tend towards pure heat. This is the underlying meaning of the second law of thermodynamics and it's a critical foundation of physics. It also means that, in the far future, everything in the universe will be converted to pure heat.

It was the Scottish physicist William Thomson (later Lord Kelvin) and the German physicist Rudolf Clausius who first (independently) realised that heat can never naturally flow from a cold body to a hot body, unless we force it by providing additional energy, for example by using an air conditioner. For the same reason, the theoretical maximum efficiency of any heat engine can never be higher than 75%, which means that at least 25% of heat is always lost. In practice electricity-producing thermal power stations never reach beyond about 55% efficiency, while the best commercial internal combustion engines used for vehicles are at about 35% efficiency, with the remaining 65% lost as heat to the environment.

Interestingly, the second law is the only law in the whole of physics where there is a clear "direction" (hot always flows to colder places, or equivalently, "entropy always increases"). All other laws of physics are independent of direction. For this reason, there is currently a new theoretical effort to relate entropy to time, since time also seems to progress in only one direction, from the past to the future. Extreme fame and a Nobel Prize guaranteed to whoever succeeds.

[4] Relative humidity is the amount of water in the air divided by the maximum amount air can hold at this temperature. See also the discussion on the heat capacity of water below.

When we are cooking, lots of the heat we pump into the kitchen is absorbed by the walls and everything else inside the kitchen, and only a relatively small part of what we pay for is used for cooking. It is a fact that there is a lot of wasted energy in an ordinary kitchen, especially if we use gas hobs, and that's why we are always looking for ways to reduce such waste. For instance, it would be great to replace my old electric range with inductively heated hobs since they hardly waste any heat to the surroundings, although they have some limitations too, as we shall see later. I might do it when my hard-working stove eventually packs up.

Thermodynamics from quantum physics?

In his 1905 paper on the quantisation of light (based on the earlier idea of Max Planck), Albert Einstein started from the premise that there should be consistency between thermodynamics and electromagnetism. That's how he proved the quantisation of light and the existence of photons as the quanta of light energy.

Recently, assuming that thermodynamics should also be consistent with quantum physics, theoretical physicists have been trying to work out whether (and how) thermodynamics might emerge from quantum physics, since quantum information (on atomic vibrations, energy states, and other quantum phenomena) also inexorably increases.

If they manage to do this, it would be extremely exciting, as it would provide a way to "unite" quantum physics with classical physics. More Nobel Prize potential.

But I digress. Next, there is the hugely important second law of thermodynamics, which states that all types of energy eventually convert to heat energy in the environment, even against all our best attempts. Since heat that has spread into the environment cannot (easily) be recovered, this law means that, in any system, all energy will gradually convert to heat, and that heat always flows from a hot body to a cold body and never the other way around. Putting a pot of cold water on an electric hot plate will cause a flow of heat from the hob to the water, and very soon the top surface of the hot plate, the pot, and the water will all be at the same temperature, as long as there is a perfect contact with the plate. If we remove the pot without switching off the electricity, the top surface of the hob will immediately start heating up and might be damaged, since there is nowhere for the energy to flow to.

While the hob is on, energy continues to flow from the hob to the pot and then to the water until they are all in a "dynamic" thermal equilibrium. This means that while the hob is on, the energy provided by the hob is continuously lost as the heat energy of the vapour molecules leaving the boiling water. If we switch off the hob, by the zeroth law, the system will go on losing energy to its cooler surroundings and will eventually reach a thermal equilibrium with, i.e., be at the same temperature as, the rest of the kitchen.

Finally, there is a Third law of thermodynamics, which says that we can't keep removing heat from a system and decreasing its temperature forever. There is a minimum temperature at which all atomic vibrations will cease. This is called "absolute zero," at about − 273.15 °C.[5] But because it is impossible to get all vibrations to stop completely in any material, we can never reach absolute zero—it is actually physically impossible! Physicists can get really close to it (closer than a billionth of a degree above it) and all sorts of crazy quantum physics happens there, but that's another very long, very strange, and very beautiful quantum physics story.

The birth of energy and everything else

All energy in the universe originated during the very creation of space–time, at the "Big Bang", about 13.7 billion years ago. But if we stay within our Solar System for now, and since it is a more or less closed system (there is almost no energy exchange with any other body as everything else is too far away), we can say that (nearly) all energy in it originates from the fusion of hydrogen atoms that takes place in the Sun.

But our Solar System itself was born from the remnants of titanic supernovas. These were explosions of gigantic stars that formed and exploded in the early stages of the Universe. So we are all the children of supernovas.

The Sun's Gifts

The only two types of energy that go into a kitchen are gas for heating and electricity for heating and everything else. So it's worthwhile taking a minute to ask where electricity and gas (and food) come from in the first place. Of course, it depends how far back in the sequence of events you want to go, but ultimately, the answer is the same for all of these: all energy on Earth, of all types, originates from the Sun, with a very small amount coming from inside the Earth, from the hot core. It was the Sun that allowed the evolution and growth of plants many millions of years ago, and these soon covered nearly the whole surface of our planet. When they died and rotted they slowly converted (while heated in the absence of oxygen) into the gas (and petrol and

[5] On the kelvin scientific scale we say it is zero (0) kelvin.

coal) that we burn on the stove (or in electrical power stations), and it's the same Sun that still allows the growth of our food, and it's the same Sun that we use to produce electricity in many other ways. Ultimately, we are all here and kept alive because of our own star, the Sun.

But hang on, you might well ask, how does the energy get from the Sun to the kitchen? To answer this we must first answer another question: how does the Sun's energy reach us here on Earth in the first place, and how does this energy move around the world?

The Sun's energy reaches the Earth by radiation. This is one of the ways that energy can be transferred between bodies.[6] In other words, the light from the Sun deposits its energy on all the objects it meets. The energy transferred from this light either increases the kinetic energy of the atoms it meets (heating the body) or produces some other work, such as the growth of a plant,[7] or both. So, how does this work?

When we say the Sun sends us heat energy by radiation, what we mean is that the Sun produces and sends out light made up of lots of little waves or particles of light, full of different amounts of energy. We call these "photons." When they reach us, they bump into our skin molecules and transfer their energy to them. These molecules then vibrate more vigorously, making us feel warm. This is radiative heat transfer to our skin. Sometimes it goes too far and we develop burns. In fact, the walls of our cells can be damaged and dry up, and actually broil and change chemical structure, but that's another physical phenomenon. We'll talk about broiling later.

Energy of light

All light photons on the electromagnetic spectrum are quantum particles, so they can behave as both particles and waves. This is known as the "duality principle," as proved by Einstein. They do not have any mass and always travel at the speed of light, c, about 300,000 km/s in vacuum or still air. As waves, they have a frequency f and a wavelength λ which are related by $c = f\lambda$.

Since c is constant, a shorter wavelength (λ) means a higher frequency (f). From Planck's and Einstein's work, we also know that higher frequency means higher energy. So, radio waves have the lowest frequency and energy while, at the other end of the spectrum, gamma (γ)-rays have the highest frequency and energy. This means that radio waves have the largest wavelength while γ-rays have the smallest.

[6] The other ways are convection, which occurs when hot fluids rise in a cooler environment, and conduction, when two bodies touch one another and exchange energy directly—see below.

[7] Much of the energy of the photons goes towards increasing the energy of the electrons around the atoms making up the plant's leaves. This is then used to drive photosynthesis and the other amazing activities that occur inside their cells.

It is the same radiation energy of the photons that is absorbed by plants and gives them the energy they need to convert carbon dioxide and water to sugars, thus allowing them to grow.

The Sun's photons also share their energy with the air molecules and atoms. They do that by hitting the atoms of oxygen and nitrogen in the atmosphere and making them vibrate and move faster, in other words by heating the atmosphere. When the photons and hotter air molecules and atoms reach the sea, they give up their energy to water molecules by the same method, that is, hitting them and exchanging their kinetic and vibrational energy. This heating of air and water is what generates the weather we have on Earth and eventually leads to rain and winds. This is why too much excess energy has huge implications for our climate.

So far so good. But let's dig a little deeper. The Sun radiates photons with a huge range of different energies. The figure below shows the full electromagnetic (EM) spectrum of all possible photon energies. The Sun radiates photons over nearly all regions of this spectrum, but few of these actually reach the earth's surface, thankfully.

The figure gives the wavelength of the various waves (in nanometres).

The photons sent by the Sun range from weak radio waves (indeed, the Sun sends us TV and radio waves too, but they never reach the Earth's surface) through microwaves (MW), infrared (IR) and visible light (the narrow bit in the middle), all the way to higher energy ultraviolet (UV) and dangerous X and gamma rays.

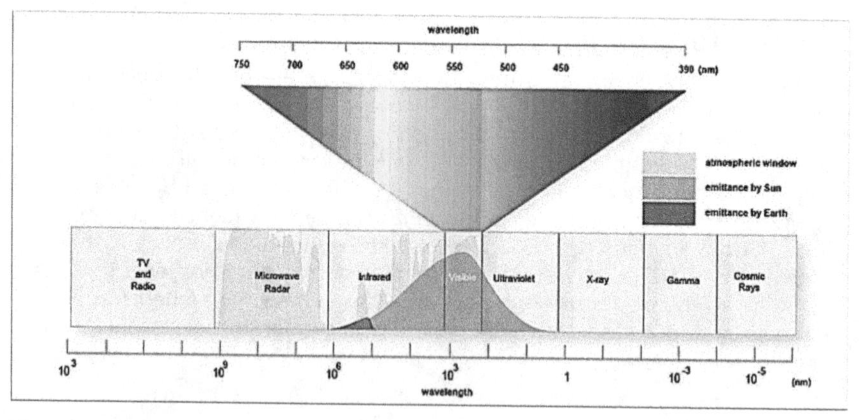

The electromagnetic spectrum, showing the visible part above and the atmospheric window and the Sun's and Earth's emittance curves below

Thankfully, the Earth's atmosphere shields the surface of the Earth from all the high energy photons (except a small part in the middle of the UV portion), but allows through most of the visible and many of the UV and IR photons, as well as most of the MW photons. This is just as well since IR photons give us heat while microwaves allow long distance communications, for example with spacecraft. Notice also that the visible part that our eyes can see is just a tiny fraction of the total spectrum, between the IR and UV parts.

The UV rays that the atmosphere lets through are split into a few UV-A, which have lower energy are relatively safe and can be seen by some animals, and UV-B which have a slightly higher energy and are potentially dangerous. These are the rays that give us the tanned look on the beach, but they can also burn through the skin when there is too much exposure. Fortunately, the even higher energy, dangerous UV-C rays are completely absorbed by atmospheric ozone, otherwise there wouldn't be any life here on Earth. However, we do use them, for example, for sterilising medical instruments.

I mentioned that IR rays from the Sun give us warmth, and that's why they are also called heat rays—they are invisible to us but we feel their heating effect.

Burning Sun

The Earth's surface receives from the Sun a very large amount of energy (mostly visible and IR wavelengths), varying from about 500 W/m^2 at mid-latitudes to over 2000 W/m^2 at the equator. On average, the amount received is about 1000 W/m^2 (at sea level), which means 30,000 J (30 kJ) of energy every 30 s on every square metre. Since the heat capacity of water is about 4200 J/kgK, these 30 kJ would increase the temperature of a small cup (100 ml) of water by 70 °C, enough for a hot cup of tea.

Of the total amount received, only about 3% is in the form of type UV-A and UV-B rays, and thankfully none of the higher energy type UV-C rays. This is because of the protective effect of the ozone (O_3) layer at the upper levels of the atmosphere.

The amount of radiation energy the Earth receives from the Sun is huge. At the height of summer, the Earth receives on average enough energy on every square meter of surface to bring a small glass of water to the boil in just 30 s. On the equator, the amount of energy received is even greater. The total radiation energy received by the Earth in a day is thousands of times the total power[8] produced by all power stations in the world in a day. In fact, it can be calculated that our global power requirements could be fully satisfied if

[8] Power is energy per unit time and is given in joules per second or watts (J/s).

we used solar photovoltaic panels with a total area of just 0.3% of the Earth's total surface area, or about the area of Sweden.[9]

It's not just the Sun that radiates IR light. All objects radiate heat in the IR region, even if we can't see it because our eyes have evolved to detect only visible light (with wavelengths between about 400 and 800 nm). When we're baking, the oven door radiates out a good deal of the heat inside and we can feel it by standing in front. A cup of hot coffee does the same. And our steaks are grilled (broiled) by irradiating them with IR rays emanating from the heating elements inside the oven. We can't see this radiation. It can be detected by some animals, although not by cats or dogs as many believe. Some old mobile phone cameras could also "see" the IR radiation as a pink glow, because they lacked IR filters. The whole Earth continually radiates excess heat to space in the IR region, and this helps to cool it down at night and keep its overall average temperature constant (see box).

By the way, inside the Earth's atmosphere, radiation is rather an ineffective way for energy to be passed around and for bodies to lose (or gain) heat. Because there is air everywhere, convection (bulk motion of air molecules) provides a more effective route, something which we'll discuss later when we talk about heat flow. Radiation also depends very much on the kind of surface of the warm material, and we use a property called "emissivity" to describe it.[10] This ranges between 0 (no radiation) and 1. Bodies that are very shiny and reflective have low emissivity while matt surfaces have emissivity approaching 1 and radiate very well. Interestingly, the best emitters are also the best absorbers of heat radiated by bodies around them. You don't need to wear dark clothes in winter, but it does help if they are shiny to make sure you don't lose your heat to the colder surroundings. In summer, the situation is more complicated. If it's particularly hot, you want light-coloured clothes, both to reflect radiation from the warmer surroundings and to radiate your heat away whenever possible. Under mild summer conditions, you want to radiate your own heat away, so even darker, non-reflective clothes are fine.

Since we just introduced the whole EM spectrum, it's worth mentioning that us, the world, and everything in it are bathed in radiation of many different types all the time. We live in a "soup" of EM radiation, from the relatively low-frequency radio waves used for TV and radio broadcasts, to the microwaves[11] emitted by wireless telephones, wi-fi routers, and satellite TV.

[9] Carbon Tracker Report, 2021.

[10] Radiation is described by the Stefan–Boltzmann law and is proportional to the emissivity of the body.

[11] Also used in radar and radio telescopes, and in our domestic MW ovens, of course.

We are of course also bathed in IR from all the objects around us, and in the daytime, by IR, visible rays, and some UV rays from the Sun too. All these rays are perfectly safe since they lie almost wholly in the visible or low-energy (higher-wavelength) regions of the spectrum. I don't know why microwaves have such a bad reputation to some people, but they certainly are much safer than the Sun's UV!

Specifically in the kitchen, we bathe in lots of IR radiation from the hot stove, pots, or hobs, and we also receive radiation from the TV we watch, the computer screen, the light above our heads, reflected sunlight, and microwaves from the wi-fi router, not to mention a tiny amount of gamma rays from certain old-fashioned night-visibility clocks and watches.

Finally, one might ask: where does the Sun get all that energy from? Well, this is where nuclear physics comes in and the answer is "by fusing together hydrogen atoms to form helium", a process called "fusion." This goes way beyond our subject matter here in the kitchen. But it is one of the ways we hope to produce energy in the future and we'll talk a little more about it later.

Climatic disturbance

Excess carbon dioxide, methane, and other molecules mostly produced by human activities (directly or indirectly) tend to obstruct the Earth's natural IR radiation of heat energy back into space, thereby disturbing the atmospheric heat balance. In other words, we radiate less to space than we receive, which gradually leads to global heating, on average.

This doesn't mean that we'll all experience warmer temperatures. The interconnectedness of the Earth's air masses and their sensitivity to even slight changes in air temperature means that the weather will gradually become more violent and unpredictable, as we are already experiencing in many regions of the world.

Chameleon Energy

We now know how the Sun's energy reaches us, so let's see how it gets converted into gas or electricity to use in the kitchen.

A whole series of events and processes leads to the production of either gas or electricity. In the case of natural gas,[12] the sequence of events is easy to picture: plants grow by receiving energy from the Sun (by radiation). When

[12] Other types of gaseous fuels are used in the kitchen, as we'll discuss later.

they die, they rot and produce methane gas (the main constituent of natural gas), which is extracted, collected, and piped to our kitchen where it is burnt to produce heat for cooking. It is a simple enough route, but unfortunately causes pollution, because it produces huge amounts of climate-damaging carbon dioxide and leaked methane, as well as other dangerous emissions. While it is quick and convenient, it is also very uncomfortable in the kitchen and very wasteful, as I'll explain later.

Electricity directly from heat

Electricity can be generated directly from heat (or more correctly, from a thermal difference) by exploiting thermoelectric (TE) materials, but the method is only used in special cases, because it's so inefficient at present if we have to supply the heat. Special thermoelectric metallic alloys convert a difference in temperature between their two sides directly to electricity by the Seebeck effect. These are the same as the thermocouples used for measuring temperature. By combining many such thermocouples, they can be used in space or planetary probes (called radioisotope thermoelectric generators, RTG) to produce electricity when far away from the Sun (e.g., on the Voyager and Pioneer probes that recently crossed over into interstellar space). The hot side of the TE is placed against a small, hot, radioactive nuclear isotope (usually plutonium 238) while the cold side is exposed to the cold space.

The same principle can also be used in the reverse mode to make small fridges, but the heat on the hot side needs to be continuously dissipated away. The net efficiency is less that 15% that of normal fridges. There is a lot of research going on to find new, more efficient TE materials and hopefully they'll be developed over the next few years.

In the case of electricity, there are a number of different routes to produce it and they all rely on energy conversion from one type of energy to another and eventually to electrical energy. The simplest of all is the direct conversion of the Sun's radiation to electricity by solar photovoltaic panels. The physics behind the conversion is complicated (it relies on quantum mechanical phenomena, just like most electronics), but the actual route is straightforward. Once it is produced, electricity is simply "piped" (along wires) to the kitchen. This is the only direct method that is both widely used and does not require a generator. Actually, thermoelectricity is another direct method, but it is still used only in niche applications because of its low efficiency.

"Electricity piped along wires" needs an explanation. Electricity is just the wholesale movement of electrons along a wire,[13] from a storage source with an excess of electrons, such as a battery, to a machine or device which uses up these electrons to carry out some work. This is what is meant by a "potential difference" and this is what is supplied by a generator.

Energy storage

There are many ways of storing potential energy and all involve placing a body under a constant force. For example, water in a dam pushes against gravity, compressed air pushes against the walls of its container, and a taut spring pushes against its own atomic bonds. Kinetic energy can also be stored in a rotating flywheel. All these methods can be used to provide energy to drive generators to produce electricity.

Electric batteries store electrons directly by placing them under an "electromotive" force produced by the potential difference across its ends as the electrons in the chemical move towards one pole. Older kinds of batteries do this by using liquid or solid chemicals charged with excess electrons. Modern batteries achieve it by using small positive ions like lithium in Li-MH or LiMO batteries, which gradually move in the opposite direction to the electrons, forcing them to move towards the opposite pole.

Electrons can also be stored directly in a capacitor where they are forced to congregate against one pole. However, in contrast with a battery, the electrons tend to be released in a burst when connected to a circuit.

Solar photons can give up their energy when they hit a solar photo-voltaic panel, but only a part of their energy is converted directly to electricity. Most of the photons' energy is converted to heat, so solar panels get very hot in the Sun and their efficiency drops significantly. This and the problem with the solar panels getting dirty or dusty are particularly thorny issues that have led to many innovations in an effort to find solutions.

By the way, the efficiency of any energy conversion method is given by the ratio of the amount of energy obtained (as electricity) divided by the amount of energy given to the system. At the moment, the best solar panels are about 25% efficient under optimum conditions.[14] In practice, however, efficiency is rarely higher than 15%. All the remainder of the Sun's energy impinging on the solar panel is lost as heat to the surroundings.

[13] Electrons actually move very sluggishly, just a few mm per second, but electricity appears to be instantaneous (at about half the speed of light!) because each electron "pushes" the one next to it in the wire, like falling dominoes. Well, this is a simplification, but not too far out.

[14] For a panel temperature of about 25 °C, with no dust and the Sun vertical in the sky.

"Nature abhors a vacuum"

Aristotle had it right when he wrote 2300 years ago that "nature abhors a vacuum," and nature will make sure that something will soon replace anything that vacates a particular niche.

This simple premise is the basis of many physical phenomena, from the generation of wind to the diffusion of atoms in materials, and even, figuratively speaking, when heat moves from a hot to a cold place.

All other routes for producing electricity are indirect methods in that they require a generator. This is simply a normal electric motor connected backwards, which means that by rotating the shaft we produce an electric current. In a generator, kinetic energy (of a rotating armature) is converted into electrical energy by "electromagnetic induction," as we shall see later. Most electric motors can be used to generate electricity this way if connected the other way round. I have one from a hair drier which produces about 12 W of power, enough to power a few LED lights, or charge my telephone, if I rotate it fast enough, and I use it to demonstrate it to students. So, to generate electricity, all we need to produce electricity is an independent way to rotate the shaft. All cars and trucks and trains have at least one generator which is rotated by the engine (in the case of petrol or diesel engines) or by the wheel shafts (in the case of some electric vehicles when they brake) and whose main function is to recharge the battery as we move. The efficiency of the best generators is actually very high, up to 60% in some cases.

One of the most useful and widespread methods for generating loads of electricity is a windmill. Wind itself is generated by localized heating of air in different parts of the world. As air is heated by the Sun, it rises (due to buoyancy, just like a cork in water, because it is less dense than its surroundings) and cold air immediately moves underneath to replace it, thereby creating wind. In some places, especially on open plains or mountains, wind blows almost all the time and windmills can work continuously. This in turn rotates the shafts of their generators, which produce electricity. This involves multiple energy conversions: solar energy heats up the air which creates wind (with lots of kinetic energy); this causes the windmill blades to produce a mechanical rotation of the generator which produces electricity. The problem with multiple conversions is that the net efficiency from start to finish is given by the product of the individual efficiencies for each conversion. In this particular case, the total efficiency (from Sun to electricity) comes out at less than 10%! Not very good at all, but at least the wind is free and the wind is renewable.

Photo-voltaic solar panels and wind-driven generators aren't the only commercially available renewable methods for generating electricity. Water falling from a dammed or a natural lake—created of course by rain from clouds which are produced by the Sun evaporating sea water—is used to drive generators (gravity-fed hydropower). Electricity may also be generated by sea waves—produced by the Sun's action again—and this method is also fully renewable and currently under development. Even the regular movement of tides—produced by the gravitational effect of the Moon—is also being developed as a renewable method to produce electricity.[15] At present, there are no other renewable methods at the commercial or pre-commercial level.[16] There are other, relatively minor renewable methods too, such as compressed air balloons at the bottom of the sea, etc., all at various experimental stages.

All the remaining methods for producing electricity are non-renewable, meaning that we need to burn some non-renewable fossil fuel, such as gas, oil, or coal, all of which are polluting and affect the climate by the release of carbon many millions of years old in the form of carbon dioxide. In these methods—which are still unfortunately being used to produce the majority of the world's electricity—the fuel is burnt by oxidising with oxygen and the heat is used to heat water. This creates steam under very high pressures in a turbine, which turns a generator. The efficiency of these methods is also high, reaching 50% or more in certain cases.

The Sun is showing us the way, again

The Sun is a boiling cauldron of thermonuclear reactions, where hydrogen atoms (with just one proton in their nucleus) fuse continuously to form helium atoms (with two protons in their nucleus), releasing humongous amounts of energy in the form of highly energetic neutrons. Just imagine billions of large "hydrogen bombs" going off continuously.

For many years scientists have been trying to find ways to control such fusion reactions in order to utilise the heat generated as a way to generate electricity on earth. Currently, the first large-scale Thermonuclear Experimental Reactor (ITER) is being constructed in the south of France, where large-scale tests will be carried out in order to enable the later construction of full-sized commercial reactors. It is an almost fully sustainable method as it uses just hydrogen as fuel (from sea water, an almost unlimited resource), but produces no atmospheric pollution and very little radioactive waste.

[15] Yes, this is also an effect of the Sun. The whole Solar System is controlled by the gravitational attraction of the Sun, the Earth and the Moon included of course.

[16] The sharp reader might object at this point and claim that in fact "there are no renewable methods at all" since even the Sun's fuel (hydrogen) is slowly being exhausted. Well, yes, but within our lifetime and even that of the Earth, this can be safely ignored.

Finally, nuclear fission power stations also use the heat generated by the radioactive decay (splitting) of uranium or plutonium fuel in the heart of the power station to heat water, to produce high pressure steam that can drive a generator. In this case, thankfully, there are no carbon dioxide emissions, but the spent fuel is radioactive and needs to be stored or re-generated somehow. In this case the efficiency is also high, pushing 50% in some of the more advanced designs.

And that's it. There are no other methods for producing electricity at present, although, as I said, there are a few new ones under development. Thinking about it, these are actually quite a few methods, aren't there? Human ingenuity at work! Who knows what fresh ideas may be developed in the future. In the meantime, let's hope that the fossil-fuel dependent methods will soon be replaced by renewable routes, otherwise the climate emergency will become a climate disaster affecting the whole world.

So, this is how we produce energy. Let's now see how we use it at home and in the kitchen to heat and cook.

Atoms Never Die

So, energy has reached the kitchen. Because we need to cook, we use this energy to heat up our food. But what does "heat up" actually mean when we are in the kitchen? How does gas and electricity give us heat?

Well, as I mentioned above, heat energy is produced (released) by two methods in the kitchen: either by combustion of a fuel (whether natural gas or bottled gas) or by forcing electrons to pass through special wires ("resistive elements") that produce heat. Let's begin with gas.

Combustion essentially means chemically reacting (at high temperature) a fuel with the oxygen of the atmosphere (an oxidation reaction) to produce carbon dioxide (CO_2) and water (H_2O, steam actually), while releasing heat energy. All normal fuels contain carbon and hydrogen atoms[17] and it is these atoms that combine with the oxygen to produce CO_2 and H_2O and a bit of carbon monoxide (if there isn't enough oxygen to go around). But what happens to the atoms when they "burn"? The answer is nothing—they just change allegiances! They break away from the original molecules and, by various means, form new molecules (all involving electrostatic attraction or electron sharing) with oxygen. If you could count the numbers of atoms before and after combustion, you'd find that they are precisely the same.

[17] For example, methane CH_4 has one carbon atom with four hydrogen atoms attached.

During a fire, most of the molecules in the fuel break up and connect with oxygen to form CO_2 and H_2O, but some go up in smoke, literally, which is unreacted carbon, and the remainder form ash, which is a waste product made of a few other types of oxide molecules (organic and inorganic).[18] The important point is that none of the original atoms are lost.

Heat can also be produced (in reality released) by combustion of any organic fuels that contain carbon (wood, paper, etc.), and we now know this to be the main source of greenhouse emissions which has led to the climate emergency.

In fact, most materials can combine with oxygen to release energy. Nearly all metals, most organics, and lots of gases and liquids, etc., are able to react with oxygen and release heat energy. When they do, they produce lots of gases and leave behind various solid oxides and other wastes. For example, burning the metals aluminium or zinc or magnesium produces a metal oxide in the form of a white powder. Most such oxidation takes place very gradually and the heat released is dissipated into the environment. Even the formation of rust from the oxidation of iron produces a tiny amount of heat. Many metals, if finely cut, can also burn violently, developing extremely high temperature flames. This was used decades ago in the form of magnesium ribbons for photographic flash "bulbs".

While the reaction products are still hot, their energy content is still very high and their atoms will continue to vibrate violently until they cool down. They also produce energetic photons, which is the light we see from a fire. In turn, these photons hit against everything around the fire, including the air, thereby transferring their energy and warming up their surroundings by radiation.

By the way, an interesting fact is that a fire can only be sustained in a gaseous environment, never between liquids or solids. The reason is that only freely interacting molecules can break up and re-react to form new compounds with oxygen. This means that, to start a fire, solids and liquids must first be converted to gases (their molecules break up and reform to give gases), which have many free atoms able to react with oxygen. They continue burning by absorbing energy from the fire around them. That's why you only need a starting flame: as the fire grows, more flammable gases are produced and this increases the flames in a positive feedback.

Let's dig deeper. How is heat *actually* produced by combustion?

The heat produced simply corresponds to the total energy before the oxidation reaction minus the total energy of the system after the reaction. This

[18] Organics are molecules that contain carbon—all the rest are called inorganic.

involves the relative strength of the atomic bonds between atoms in the molecules before and after combustion. In the case of oxidation of fuels, the total bonding energy between the atoms in the fuel (carbon–carbon, oxygen-carbon, etc.) happens to be greater than the total energy between all the atoms in the resulting molecules. So, when molecules are broken apart by the input of a starting flame (in some devices, there is a pilot flame that triggers the combustion), they prefer to react with oxygen because the energy needed to form the new molecules is lower. The difference in energy is released as heat, producing strongly vibrating atoms in the combustion products.

The colours we see in any flame are the result of a basic quantum mechanical phenomenon. Electrons around atoms occupy different energy levels (the orbitals mentioned earlier). In fact, it is a basic premise of quantum mechanics that electrons cannot have any other energies than those allowed by the orbitals they occupy. Some of the heat energy in a gas flame is absorbed by electrons that can then "jump" momentarily to higher energy levels. When they "fall" back to their normal ("ground") energy state, they emit the energy difference as photons of light with various frequencies, sometimes within the visible portion of the spectrum. This is how all electric lights work, from the old incandescent lights, to fluorescent tubes and the modern LED (light-emitting diodes). All materials have their own characteristic energy levels, so they emit different colour light depending on the difference between the energy of their ground state and the higher level energy the electron has reached by gaining sufficient energy to make that jump. For example, nitrogen or sodium in a flame emits yellow light, lithium emits red, and copper emits a nice turquoise light. One of the most beautiful quantum mechanical demonstrations I use for teaching is observing the flame colours produced when we throw some metal salt in a small flame.

In a gas hob, two colours are generally seen. In natural gas flames (made up mostly of methane), most of the flame should be blue because of the excitation of the electrons in the carbon dioxide produced by complete combustion, as well as any remaining oxygen. In bottled gas (LPG: liquefied petroleum gas), the tip of the flame is slightly yellow or orange because the flame burns at a higher temperature and excites the electrons in the atmospheric nitrogen. Under ideal conditions, meaning sufficient oxygen and no other organic molecules (e.g., from a dirty hob), no other colours should be visible.[19]

It's worth remembering that all bodies emit some light, even those at room temperature, but since the temperature is low, the photons emitted have very

[19] If you see other colours (red or yellow), it may mean that there is not enough oxygen, perhaps due to a blockage or dirt. This can produce carbon monoxide and soot.

low energies and fall in the infrared region, which we cannot see, as we discussed before. But it is possible to see them with specially designed IR-sensitive electronics which contrast the heat of the bodies against the cooler surroundings.

If the heat released during combustion is sufficient to continue breaking up and rearranging the atoms, then we talk of an exothermic reaction. If, however, we have to keep on adding energy to keep the fire going, then we talk of an endothermic reaction.[20] Most fuel reactions with oxygen are exothermic: once they start, they can keep going on their own—the flame acts as its own pilot light and sustains the reaction. We see this in a gas hob and in devastating forest fires. On the other hand, most reactions in living bodies are endothermic and energy has to be provided continuously in order for most biological reactions to continue unabated.

Interestingly, every normal fire (from any fuel, gas, oil, wood, or coal) produces a range of gases. That's because there is often not enough oxygen to completely burn off all the fuel, and also because the surrounding temperature is never high enough to allow full oxidation in the periphery of the fire. We'll discuss such gas products later when we talk about gas fires in the kitchen and their problems.

The Dance and Flow of Heat

The main route by which heat energy moves around is transfer between materials. As I mentioned earlier, in addition to radiation, which emanates from all bodies, mostly as IR radiation, heat can also be transferred via two other means: conduction and convection. In both cases, heat is transferred when vibrating atoms gradually jostle their neighbours, causing them to vibrate more, so that they in turn jostle their own neighbours, and so on, just like dominos knocking each other over one after the other. Conduction occurs between solids or liquids when they touch (e.g., the hot hob and the bottom of a pot), while convection occurs in liquids and gases when there is wholesale movement of molecules, e.g., when boiling water or when heating a room with a radiator. Over time, heat spreads in all directions through a material because of the jostling of the atoms. A cooler atom can only gain energy from a hotter one, as the laws of thermodynamics dictate. That's why we see that heat is always conducted away from a hot region towards a less hot region. Eventually, the whole body reaches the same temperature, i.e., all its atoms

20 "Exo" in Greek means "outside" while "endo" is the prefix meaning "inside", implying that exothermic reactions exude heat while endothermic ones require additional input of heat.

will vibrate equally strongly. At the same time, the body radiates heat away and, if there is no fresh input of energy from somewhere, the original total amount of heat will be reduced, and the average temperature of the body will drop as the atoms lose energy and vibrate less and less.

This is important—heat energy is always being transferred. Heat always moves from a warmer body to another, less warm body, continuously. As we saw earlier, this is a fundamental law of nature: heat energy can never stay put in one place. Our own body loses heat by radiation to our surroundings, although we also receive radiation from our surroundings, by conduction (if we touch something cold), by convection (if we stand in a draft), and by the energy we use to move muscles, to speak, to think, etc. This means that our body needs to produce heat all the time by the metabolism of food, just to keep our temperature at a healthy 37 °C—that's why food and cooking are necessary. Even in space, energy is continuously lost by a body to its surroundings by radiation. A spaceship is always warmer than the surrounding space (which is terribly cold, at just 2.7° above absolute zero, about − 271 °C), so it always loses heat. But note that, since space has almost no atoms or molecules, conduction or convection cannot occur.

Conduction of heat occurs by direct contact between atoms in solids or molecules in fluids and is a slow process, which depends on how good the contact is between the bodies, for example, between the bottom of the pot and an electric hot plate or hob.[21] If we force the atoms closer together (for example, by exerting pressure), atomic contact improves and conduction increases. In reality, of course, no two materials ever actually touch one another because of the repulsive forces between the atoms. But the vibrations of the atoms do get through from atom to atom. By pushing them together, heat conduction between them increases up to a maximum. That's why you can feel the heat of a heavy item, but not so much that of a light item, even if they are made of the same material and are at the same temperature.

When you cook, have you noticed that the pan or pot handle gets gradually hotter, even if it is made of some non-metallic material? Heat conduction is the reason, and the handle will go on getting hotter even after you finish cooking, as the total heat is distributed throughout the pan, until the whole thing reaches a constant temperature. At the same time, heat is lost to the surroundings by radiation (in every direction) and conduction (to the now cooler hob), as well as by some convection, until the pan and its contents lose all their excess heat and cool down. This can take a long time, depending on the "heat capacity" of the system as a whole. This is the amount of heat

[21] Heat conduction through solids is described by Fourier's law and is proportional to the temperature difference and the cross-sectional area of the conductor or the contact region.

that is contained in the pan itself and also in its contents, and it varies very widely, depending on the specific materials involved. Strictly, it is the amount of energy the body must absorb to raise its temperature by one degree. Water has one of the highest heat capacities (per unit mass) of all materials, only surpassed by ammonia, hydrogen, and helium. This is just as well as it makes it difficult for the Sun's radiation to evaporate it, so it is not easily lost to space. Water is clearly the most amazing material we know of and we'll say more about it when we look more deeply at cooking phenomena.

Heat conduction also occurs between air and its surroundings. In winter, the nicely vibrating atoms of the heated air inside the kitchen hit against the colder walls giving up some of their heat. This is then conducted through the wall to the outside, where it is radiated away to the colder surroundings. Conduction through materials is described by the "thermal conductivity" of the given material and varies widely from close to zero for very good insulating materials (bad heat conductors) such as expanded polystyrene (which contain air bubbles), to very high values for aluminium, silver, and copper (very good thermal conductors).[22] Most materials we use in the kitchen such as steel, ceramics, glasses, and plastics have moderate conductivities. Water also has a moderate conductivity, which becomes painfully obvious if we try to remove a baking tray from the oven using wet kitchen gloves. Dry gloves are fine, because they contain large amounts of air (with very low conductivity) trapped between the layers of textiles, which are also very good insulating materials. The same principle is used for padded jackets, etc.

Convection works the same way as conduction to move heat around, but only in liquids and gases. Convection is simply speeded-up conduction in the gas or liquid state, as gases and liquids are free to move around and mix, increasing the rate at which atoms can smash into each other and share their vibrational energy. You can see this when water is heated in a pot and the hot water at the bottom rises to the top. The same happens with a convection heater, although the effect is largely invisible. Air atoms come in contact with the hot heater surfaces, where they pick up heat by conduction and rise and interact with the cold air atoms, while colder air comes from below to fill the vacant space. This continues in a nice, continuous movement which gradually heats the room. But what makes the water in a pot and the warm air rise? Think back to what we said about vibrating atoms. The more they heat up, the more they vibrate and the more space they take up, like dancers jostling on a dance floor. This means that hot fluids (e.g., water or air) occupy a larger

[22] Conduction of heat through a material resembles waves spreading from a source, and the study of heat conduction uses the quantum–mechanical concept of wave-like "phonons" spreading through the material, complete with interference with other waves and reflections at interfaces, pores, and cracks, treating them therefore just like photons.

volume than colder fluids, for the same number of atoms, and are thus less dense, i.e., lighter, than when they are colder. This means that a hot fluid rises (because of buoyancy, an effect of gravity), displacing and interacting with the colder fluid above it. By doing so, the hot fluid cools down and falls once again to repeat the process and gradually heat the room or boil the water.

In a room, hot air from a heater rises to the ceiling, and if it's not well insulated, the hot atoms will transfer their energy quickly to the cold ceiling by conduction, whence it will be lost without heating the air in the room. As a result, the warmest place in any room is always near the ceiling and that's why in older houses beds were sometimes on a raised platform. That's also why the warmest flats in a block are the ones sandwiched between floors—you get energy from the flat below and you lose only a little energy to the warm flat above.

To help the atoms and molecules share their energy uniformly, stirring a fluid always helps, as it brings more molecules in contact with one another. Even in winter, using a ceiling fan can bring some of the hot air down from the ceiling to mix with the colder air below– but only if the ceiling is well insulated. That's why an air conditioner should always aim for the bottom of the room in winter—the hot air will mix nicely as it rises in the room; but always aim for the top of the room in summer, since we want the cold air from the air conditioner to sink and mix around.

To speed up and increase the uniformity of heating in a pot, it helps to stir the food, but this is not always necessary—convection in the soup during heating will do the job very nicely, as long as there is sufficient water in the pot and an easy, unrestricted flow! But there are lots of other provisos about this, as we shall see later.

Shifty Atoms

Convection helps to bring molecules and atoms of liquids and gases into contact with one another so that they can exchange or share their energy. In solids, heat is conducted by pseudo-waves (the phonons mentioned earlier), but it's also possible for the atoms themselves to move through the atomic lattice of a material, aided by the vibrations of both the lattice atoms and the moving atoms. This shifting movement from one atomic position to an adjacent position is called atomic diffusion, and it occurs whenever there is a difference in concentration of the particular atoms. Atoms prefer to populate any region in the lattice where there is a dearth of their own type, until they are completely evenly distributed, at which point diffusion stops. This is a very important phenomenon and one we encounter all the time when cooking, as will become obvious in the rest of the book.

The rate of diffusion of atoms through another material (e.g., an atomic lattice) depends strongly on the temperature, the strength of the atomic bonds in the lattice,[23] the relative size of the atoms involved, the different lattice structures (interfaces between crystallites, microcracks, etc.), the difference in relative concentrations, various external influences, and other parameters. For each lattice-diffusing atom system, under specific conditions, we measure a diffusion coefficient that gives the rate of diffusion in that particular system. It's also possible to have molecular diffusion, although this is more difficult due to their larger size. Diffusion also occurs through liquids, where it is generally much faster than in solids because of the more tenuous bonds between the atoms or molecules in a liquid.

It's obvious that both molecular and atomic diffusion are directly related to cooking and many other events in the kitchen. In a pot we try to encourage diffusion between ingredients by heating, stirring, and adding various atomic bond-altering agents, as we'll see later. During baking, diffusion is slower, and in cakes or breads we want the surface of a cake to be sealed (so that all pores are closed) as soon as possible. This serves to trap the carbon dioxide produced within the cake. By sealing the cake's surface, we reduce the gas diffusion rate and keep it inside, forcing the individual pores to swell and the cake to rise. We'll discuss this in some more detail later.

Diffusion is also important when we want to distribute various delicate herbal aromas (ethereal or essential oils) through solids in a pot or casserole without damaging them. For this we wait until the vegetables or potatoes (and perhaps meat) are already infused with liquids, and then we add the herbs at high temperature and immediately switch off the hob. The herbal essential oils diffuse (or infuse) quickly into the vegetables, where they are protected until cool.

Since diffusion occurs much more quickly in liquids than in solids, timing is important when adding the various ingredients in the pot. For example, if we wait until the very end, when the soup or stew is very thick to add condiments and salt, they may not distribute as well as if we had added them slightly earlier, when the mixture is still liquid. The same is true with cake batter. Baking powder and (bicarbonate of) soda should be mixed well with

[23] For atoms to diffuse through a material, bonds between lattice atoms must stretch, but transient bonds between the diffusing atoms and the lattice atoms must also break and reform. This is a very complicated probabilistic phenomenon. It is treated statistically on the basis of Boltzmann probability theory to obtain Fick's laws of diffusion.

the flour before adding to the liquid butter-sugar-egg mixture, as this ensures good distribution and better diffusion before baking.

Diffusion is a ubiquitous process in cookery and also occurs in many other situations in the kitchen. We'll consider many diffusion situations as we go through the book.

Quantum Weirdness

Closing this introduction to the main physical phenomena and laws that control most things that happen when cooking (and in the kitchen in general), I would like to add a few comments about quantum mechanics. As we go through the book, there will be many occasions where we'll refer to quantum mechanics (QM) as the basis of some of the phenomena we encounter in the kitchen. We've already encountered some.

Although in most cases we are sure about those attributions to QM, in other cases we are not so sure or we are still working on them. After all, that's what scientific research is all about. Testing and re-testing ideas to make sure they remain correct and relevant under new situations we discover.

Uncertainty at the heart of matter

The discovery (and repeated confirmation) of quantum mechanics at the beginning of the twentieth century and the discovery of the probabilistic nature of the properties of atomic particles has upended the certainty of classical physics, at least at the atomic level. A major pillar of QM is the "uncertainty principle" discovered by Werner Heisenberg, which states that you can only know accurately either the position of a particle or its velocity (momentum), but not both at the same time.

The probabilities of the properties of any particle (its "state") are described by its "wave function", developed by Erwin Schrödinger, which shows that such a particle exists in all possible states simultaneously, something which is called "superposition of states". Only when we perform a measurement do we get particular values for the properties we are measuring, and we call this "the collapse of the wave function".

Quantum mechanics only describes the properties of atomic-level particles and it cannot be used to describe larger-sized systems. This is the reason for the confusion surrounding "Schrödinger's Cat" thought experiment, which should not be taken literally.

Quantum mechanical explanations and mathematical formulations of quantum mechanics work, and they work very well. Literally hundreds of discoveries and inventions are based on QM, and every time we test it, we find that its predictions are correct. But if you ask me, or any physicist, whether we fully understand the fundamental physical basis of QM, well, I'd be lying if I said we fully understand how it works. Even Richard Feynman, one of the great physics minds of the second half of the twentieth century and probably the foremost expert on QM in his day, has been quoted as commenting: "I think I can safely say that nobody really understands quantum mechanics"! It's also well known that Einstein himself (one of the founders of QM) was famously doubtful about the fact that QM deals only in probabilities, not certainties, and said at one point: "God doesn't play dice with the Universe". What he was concerned about was the discovery that the properties of atomic and sub-atomic particles can only be defined as having probabilities, and it is only when we carry out an experiment to measure a property that it actually assumes a specific value. This probabilistic nature of sub-atomic particles means that they can have any properties they want when "we are not watching them". This is what bothered Einstein, but it has turned out to be true every time. Because of that, he was sure (as many physicists still believe) that QM is just the tip of an iceberg, and a more general physical foundation is waiting to be revealed. Until we discover this all-encompassing physics, we continue working with the quantum rules we have discovered, which do their job excellently.

But the probabilistic nature of quantum particles also means that, occasionally, a sub-atomic particle can take a value (energy, position, etc.) that, by classical physics, it shouldn't be able to. We call this the "quantum tunnelling" effect and it is probably the most significant of all quantum mechanical effects, because it explains many everyday phenomena. It seems to explain how catalysts and enzymes work in living things, but it also explains why the Sun shines and it provides the basis for many discoveries and inventions in electronics and laser physics.

It's really weird. Considering its incredible impact on our lives, QM is probably the most successful piece of physics ever discovered, and it gives us amazing applications (many in the kitchen and in the pot, as we'll see), and yet we still don't fully understand how it works.

But the rules of quantum mechanics work, and they work very, very well. There is not the slightest doubt that they work and we apply them everywhere with full confidence. Our modern civilization, with its computers, smartphones, lasers, advanced materials and medicines, the internet, LED lights, solar panels, etc., etc., is firmly based on the weirdness of quantum mechanics!

Anyway, without further delay, let's commence our investigation of the physical phenomena in the kitchen, starting by looking at some of the edible food cornucopia we use and the physical principles that are hidden in them, some of them quantum mechanical. Afterwards, we'll look into the physical (and chemical) phenomena that happen when foods are cooked, and consider their products, both tasty and otherwise.

3

A Physics Cornucopia

"Food is simply sunlight in cold storage"

John Harvey Kellogg

G. Vekinis, *Physics in the Kitchen*, Copernicus Books,
https://doi.org/10.1007/978-3-031-34407-7_3

Cooking, apart from anything else, is simply mixing ingredients (sometimes in a specific sequence) and heating them carefully in order to enable inter-diffusion of structures and molecules, as well as the partial breakdown of cellulose and other molecules to enable easier digestion of nutrients. Let's first see what physical phenomena exist in these raw ingredients, starting with the most basic: water.

It sounds very simple and easy, doesn't it? Well, yes it is, if all you want is to survive. But if you also seek pleasure, then cooking becomes a bit of an art. And like all art, the more you understand your medium and your methods, the better the result.

Cooking involves a myriad combinations of hundreds of ingredients heated using many different methods. Every one of these combinations gives a different result which itself depends on how it's prepared. The nature of the ingredients we use is paramount, since their molecular and structural properties are affected by the cooking parameters that we use, such as temperature, time, oxygen, and motion (mixing, etc.), all of which control the physics (and the chemistry) behind cooking. That's why no cooking result is ever exactly the same. There are too many combinations!

In this part of the book we'll look at some of the salient physics hidden inside the basic ingredients: meat, fruit, vegetables, salt, sugar, etc. And we'll start with, what else, the most basic ingredient of all: water.

Water, the Beginning of All Things

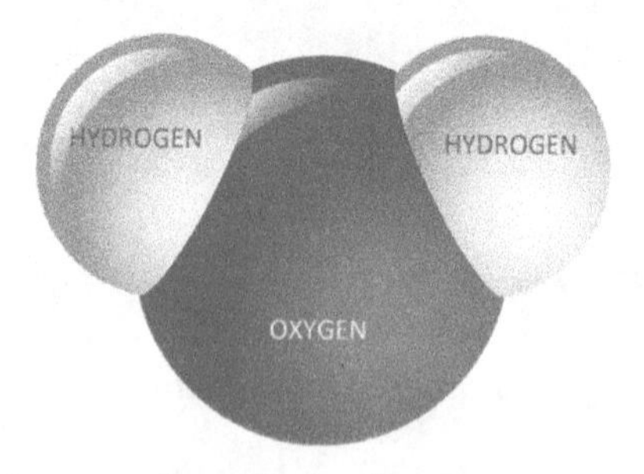

A molecule of water

There is one thing that is never missing when we are thinking of cooking: water or H_2O. A simple molecule made up of one oxygen and two hydrogen atoms that makes life and cookery possible. It's true, all life on Earth depends totally on the availability of water. There is life without oxygen, without sunlight, without heating, without air. But there is no life without water.

Everyone knows that the human body (and that of most animals) consists of up to 75% water, most of it inside the trillions of cells that make up our muscles, our organs, and our bones. Without it, none of those cells would stay alive and none of the processes that keep us alive would be possible. Water makes up nearly 90% of the blood in our bodies and about the same amount of our brain. Think of it. All our thinking capabilities reside in a convoluted blob that is 90% water. I'm writing this while my brain cells are just swimming in water! Like most liquids in the body, blood is mostly water so every beat of the heart is almost immediately felt[1] at the very tips of our toes or fingers and enables the intricate system of blood circulation. Transfer of nutrients across our cell walls occurs by diffusion, and this is only possible because they are dissolved in water. And water carries the oxygen-laden blood cells to the tiniest of capillaries everywhere. Every day we lose lots of water (up to 2 L a day in the summer) by perspiration and natural waste processes, so it is critical that we replenish it by drinking and eating. Water's special properties make life possible on Earth and its very special properties make me think that this observation must be valid throughout the universe.

Diagnostic water

When we undergo magnetic resonance imaging (MRI) to enable a diagnosis of an internal problem, we utilise the fact that a certain quantum mechanical property, called "spin," of the hydrogen atoms in water molecules and also in many lipids is affected by externally applied magnetic fields.

Because water is everywhere in the body, an MRI scan gives very clear pictures of our internal organs, and a doctor can then assess any problems. For example, inflammations are usually accompanied by high concentrations of water (e.g., in swellings) and they can thus be pin-pointed easily.

High speed MRI scans of the brain, the heart, and other active organs also help to understand their operation and the functionality of their various substructures.

Water is an almost universal solvent, i.e., it can dissolve nearly everything, including metals and ceramics. It does this because it is a polar molecule: it is triangular, with the H–O–H angle being about 105° (see figure). The oxygen side is negatively charged and the open hydrogen side is positive. Being polar it acts as a dissociation agent for most other materials around it, because it has the electrical energy to break up other molecules.

[1] Gases are compressible, but liquids are almost completely incompressible, except when they contain dissolved gases, and transmit forces extremely fast.

The huge amount of water in the oceans contains nearly every naturally occurring element in it, including probably more gold than all the land. The same is true for platinum, iron, and nearly every other metal. It contains more salts (mainly the familiar table salt NaCl, but thousands of other salts too) than all nearby landmasses. Only some man-made polymers and some chemicals are considered insoluble in water. Or maybe we just haven't waited long enough.

But what does water actually do in food as it enters the body? Let's follow it and look at all the physics involved in its many roles.

The first thing we see is that water is the carrier for all foods. All foods are generally loaded with water, and if any food is dry our saliva makes sure it gets wet. Most nutrients are dissolved in water by breaking them up into atoms or simpler molecules and attaching them to one or the other of the water molecule's poles, although in some cases a whole lipid molecule simply attaches to a water molecule without breaking. Nutrients cannot enter the body without water and they cannot be absorbed and used by the body unless they are dissolved in or attached to water.

Water is actually very dynamic, which makes it even more active. While it stays as a complete H_2O molecule, it does its job well. However, as I mentioned before, it does not generally stay whole all the time, but continually dissociates itself into lone hydrogen ions (just bare protons) and hydroxyl (OH) (free) radicals which reform almost immediately. But while it is broken down, the OH radicals can do quite a bit of damage by reacting with and oxidising nutrients, which means they can be quite harmful. The body detects this, so it has lots of spare hydrogen ions which react with the radicals to neutralise them as soon as possible. Recently though, it has been determined that excess OH radicals also serve as sentinel signals for the body when something is out of balance (as in inflammation), so they do some good as well. At the same time, the lone hydrogen ions from water and from nutrients enter cells via the cell membrane to give up their energy.[2]

Natural rain water contains many different types of ions, mostly as salts (carbonates, chlorides, etc.), which can react with other molecules in food, giving rise to various flavours or even affecting flavours. Because of this, it isn't at all rare to notice that particular foods taste differently if made with water from different sources.

[2] The whole process is very complicated, but it is the basis of energy production in a body.

Water shouldn't exist!

The simple molecule H_2O is actually even more fascinating than indicated by all the properties mentioned. A quantum mechanical analysis of its structure shows that stable water should not even exist, as the hydrogen and oxygen atoms always seem to prefer to break up, instantaneously.

But we know that water exists and its existence is due to the Heisenberg uncertainty principle (that pillar of QM), which allows those weak hydrogen bonds to form just long enough for water to exist in its liquid state. But only just, because the bonds keep on breaking up and reforming all the time. Liquid water seems to be an illusion!

Just as well actually, because we also need water to break up easily enough to be able to dissolve other materials.

Amazing dihydrogen monoxide (H_2O)!

Water is really difficult to understand. From my own research I have observed it forming transient structures while liquid and changing the properties of reaction products. As we saw, it is one of the most difficult materials to heat and evaporate. Water has the highest heat capacity[3] of all liquids, but also the highest latent heat of vaporisation[4] of all materials. This means that you need a huge amount of energy to heat up water and when it reaches boiling point (100 °C at sea level), you need another huge amount of energy to evaporate it, while both the remaining liquid water and the resulting water vapour remain at the same temperature. And this for a molecule held together by weak hydrogen bonds that keep breaking and reforming all the time. There is more. Water also has one of the highest latent heats of melting. A large amount of energy is needed to convert solid ice to liquid water, while the whole mixture remains at 0 °C. And thank goodness for all that, otherwise there wouldn't be any life on Earth.

Water is quite unique in many other ways too, and we'll discuss many of its special properties in some detail in various places throughout this book. For example, it is the only liquid that expands when it freezes—that's why icebergs float in the sea—and causes broken pipes in winter due to this expansion. This happens because of the way the hydrogen bonds keeping the molecule together try to minimise the total energy of the structure when the atomic vibrations decrease at freezing temperatures. For the same reason, expanding ice damages tomatoes, lettuces, and cucumbers in the freezer. But not cabbages. Bear with me and we'll soon see why. Water will accompany us throughout this book.

---

[3] The amount of heat input to convert a given mass of liquid to steam. We'll say a lot more about this when discussing the way fridges work and other issues.

[4] The amount of energy required to convert all the liquid to gas.

Edible ... Plastics—The Building Blocks of Life and Food

If you thought man-made plastics were a ubiquitous man-made innovation, you'd be mistaken! Life has been making polymers since the very beginning. All proteins—the building blocks of life—are long molecules, we call them macromolecules, chains made up of thousands of repeating molecular units, just like the man-made variety. The famous double helix DNA, which holds all the information necessary to make another individual, is just another huge macromolecule, billions of molecular bases long, i.e., a protein polymer.

Later on in this book I discuss the details of man-made polymers, including how their made, their characteristics, etc., but here we'll look at polymers as food—because most foods are actually polymers. Polymerisation is of course chemistry, but the way the monomers join up to create chains and the ways they fold so curiously and repeatedly (and unerringly) are guided by physical principles, especially considerations of electrostatics and quantum mechanics.

So, first of all, let's look at the raw materials we use for food. Since life is based on proteins, it's clear that all animals and plants are made of polymer chains coiled and wrapped around each other, and continually interacting together to give life. To form such molecules we need catalysts (we call them enzymes in living things), which speed up the bonding between the monomers, i.e., the separate simple molecules or atoms. If these enzymes weren't around, it would take centuries for any protein to form spontaneously. Enzymes are chemical compounds, some of them protein chains themselves, which contain atoms of certain metals such magnesium, iron, calcium, cobalt, molybdenum, zinc, tungsten, and others which are known as the "essential trace elements" in every multi vitamin tablet. They speed up chemical reactions in all living things by millions or even billions of times, and they enable the manufacture of billions of proteins every second in every cell of your body. You might even say that it's these metal atoms that are the basis of life on Earth, and you wouldn't be far wrong. In fact, the most reasonable suggestion for the emergence of life is the "hydrothermal vents" hypothesis, which points to these vents at the bottom of the sea, where metals from deep inside the Earth work their way out into the cold ocean waters. But that's another story.

The way such catalysts work in detail is complicated (it's a quantum–mechanical phenomenon again), but basically they act as a half-way house for atoms or proto-molecules to meet one another and bond to form the protein chain. If the temperature increases, the chemical reactions accelerate even further because increased atomic vibrations facilitate easier bonding between atoms, even more so in the presence of the catalyst. And the longer you allow the reaction to take place, the longer the macromolecules will be, i.e., they'll attain a higher molecular weight and, depending on the polymer, they may become harder or more elastic and/or stronger.

In the case of plants, we find the same story with added ingredients. They are all made of proteins, cellulose, etc., all of which are polymers. Trees depend on lignin (also a polymer) for support and nutrients are brought up from the ground through polymer capillaries to the leaves and other structures, also made of polymeric materials.

The exoskeleton of prawns, crabs, and insects is made of chitin, another strong polymer. Polymerisation is nature's main tool for making living things!

Polymers are created and abound during cooking too. We'll talk about them presently, but let's have a quick taste right now. In addition to the existing polymer proteins we add in the pot to make a meal, cooking forms new polymers from the ingredients we use. For those of you that enjoy cooking, you've probably seen that mixing oil and some acidic agent, such as vinegar or lemon, or even some wine in the food and allowing the sauce to reduce by slowly boiling and whisking, you'll produce a lovely smooth, thick sauce. This is in fact an emulsion. The more you reduce it the thicker the sauce gets. You've created a liquid polymer! What happens is that the acid acts as a catalyst for the oil and water to bond slowly together to produce a complicated macromolecule. The reaction is slow, but if you do that under pressure (in a pressure cooker) the boiling temperature will increase and you'll get a thicker emulsion more quickly, i.e., the molecular weight will be higher. We'll talk about cooking polymers later, and see why the use of pressure cookers can speed up cooking.

Polymerisation during cooking is also favoured when we mix starch (itself a polymer) with water and boil it for a while. If you are boiling pasta, you can keep some of the remaining water, which now contains dissolved starch and gluten, and use it for thickening sauces quickly without much need for reduction. The short chain molecules in the water attract other molecules, increasing the molecular weight. The sauce becomes a thick polymer gravy with added taste.

Polymerisation during cooking also occurs when mixing water with sugar and a few added drops of lemon, or some brandy, or just leaving some lemon or orange seeds in it. Again, the acid or alcohol acts as a catalyst for the polymerisation reaction to produce syrup.

On the other hand, most natural polymers break down by extended cooking, making food easier to digest. If you try to chew raw leeks, or meat for that matter, you'll have a real problem, and most probably you won't be able to soften it enough to swallow it. And digestion will be very slow. But cook it a while—either on a fire or in the pot—and you'll be able to chew it and digest it very easily. The long polymer chains will have broken into short chains or even their basic building blocks, the amino acids. In fact, one of the most widely accepted theories of human brain development hinges on the huge boost in nutrient intake of early humans when they learnt how to cook! Cooking probably came before civilisation.

Physics in a Cup of Tea

Fancy a cuppa? Everybody (well, nearly everybody) likes a good hot cup of tea, especially in winter. But even in the Sahara desert, in the middle of a hot day, the Bedouin love their tea, because it cools them down. This is because it encourages increased blood flow near the skin and this increases heat loss. Tea is made by hot water forming new bonds with the surface atoms of the tea leaves (in physics we call it simply "wetting" the surface). This quickly softens the tea leaves and allows the release of many types of molecules ("catechins[5]", tannins, caffeine) from the leaves, dissolving some of them. In fact, the water molecules actually penetrate into the tea leaf cells by breaking some of the atomic bonds of the cell walls (i.e., by dissolving the walls). This allows some of the aromatic tea compounds within the cells to dissolve in the water. Once there, these compounds slowly diffuse through the water. The extent of this spreading can be aided by stirring, which makes water containing dissolved tea molecules continuously touch and interact with new neighbours. This allows for many more interactions between the water and tea molecules, enhancing the taste and colour. The diffusion of molecules continues and increases as long as we leave the tea in the hot water. The colour gradually deepens and the tea becomes "stronger," that is, more saturated with the various molecules. But at the same time, two other effects take place. One is the oxidation of the tea, by reacting with the dissolved atmospheric oxygen, darkening the tea and making it more bitter. In parallel, if the water is too hot (more than about 90 °C) many of the molecules are broken up (dissociated), drastically changing the taste to something much more bitter.

[5] The main difference between green and black tea is the extent of oxidation of these catechins in the tea leaves, which produce dark theaflavin and thearubigin.

And all these events are brought into play just by making a simple cup of tea!

By the way, if you want only the aroma of the tea and not its bitterness, switch off the kettle just before it boils and use water at about 90 °C, not 100 °C, the boiling point of water at sea level. This way you will preserve some of the aromatic substances that will otherwise evaporate along with the steam. Have you noticed how tea smells great during preparation, but that the aroma soon stops? That is the reason—most of the aromatic substances in tea evaporate easily at high temperatures. That's because their molecules are generally weakly bonded and readily leave the liquid and dissociate at high temperatures.

On the other hand, making tea with water at even lower temperatures does not give sufficient energy to dissolve the aromatic substances in the first place. In addition, the diffusion is then much slower and the tea becomes rather tasteless, with a very weak aroma. Perhaps the best tea would be on a 3000m high mountain (Olympus!) where water boils at close to 90 °C.

To avoid the bitterness of overheated tea but get the most pleasure, without worrying about the actual temperature, I use boiling water and keep the teabag in the hot water as little as possible by moving it in and out (dunking it) a few times. This activates diffusion of the aromatic molecules and nutrients but keeps the tea from overheating and the sensitive molecules from dissociating. At the same time, oxidation by the air darkens the tea very quickly without damaging it. The results are very pleasant, although I must admit taste is in the "eye" of the beholder. Some people prefer the strong, bitter "brew" with lots of dark thearubigins, others don't. There's no accounting for taste, then.

Adding lemon to tea has the interesting effect of suddenly lightening it up. If it is added soon enough after pouring in the water, it will preserve the tea aroma. This is because it increases the acidity of the tea (decreases the pH), especially in places where the water is rather alkaline (i.e., it contains a lot of salts), and this protects the aromatic molecules from alkaline damage.

Adding milk to tea can also have a significant effect. If it is skim milk (no fat), then the effect is minor, but if the milk contains lipids (fats), then they react and bind with some of the catechins in the tea, blunting their taste.

Finally, a word about those dark deposits on the inner wall of some tea cups that are well-nigh impossible to remove by washing. This is due to oxidised pigment molecules that float around in the tea and are partially responsible for its dark colour. The problem is that they diffuse into the tiny micropores and microcracks on the inner surface of the cup. High quality porcelain (with added kaolin) is almost immune to these discolourations, as it is fired at very high temperatures (over 1300 °C), which melt the surface of the ceramic and close all such pores quite effectively. We'll talk about ceramics in the kitchen later too.

Fruit and Vegetables—Catalysts of Life

The bacteria that give us life

Mitochondria exist in their thousands within all plant and animal cells and are the only structures in the body that are able to obtain energy via oxygen respiration. They enable our life in an oxidising world.

More than 50 years ago, Lynn Margulis first realised that mitochondria are actually the descendants of a free-ranging aerobic bacterium that was somehow "captured" billions of years ago by an anaerobic archaic bacterium, and their symbiotic descendants were so successful that they evolved into the present day cells. They are the only structures that are able to pass on electrons from food to "toxic" oxygen and thereby enable metabolism and produce the "energy currency of life", the molecule adenosine triphosphate (ATP). They still retain a small number of their own genes, but they rely on the cell and its own nucleus and DNA to produce proteins and themselves.

Each mitochondrial membrane houses thousands of continuously rotating nano-motors whose function is to feed through protons (hydrogen nuclei from food), thereby setting up an electric field stronger than a lightning bolt (30 million V/m) across the membrane to enable our metabolism.

It may seem odd to talk about the physics of fruit and vegetables, but physics is involved here as well. So, let's have a look.

First of all, some oily vegetables and fruit such as sunflower seeds and olives have incredibly high heat content. For example, sunflower oil and olive oil contain more energy than an equal mass of butter and much more than the same amount of meat of all types. Including them in our foods provides large amounts of energy for our own metabolism. It's not surprising that such foods have always been seen as the most valuable of all, especially in ancient times.

But even more importantly, all vegetables are excellent sources of metallic elements (absorbed from the ground by capillary action[6] as salts dissolved in water and diffusing into the plant cells). These are critical for the manufacture of enzymes, those magic catalytic "keys" that enable the production of

[6] Due to water's high surface tension.

all proteins in every living body by the metabolism of energy-rich foods. As I mentioned above, the amazing thing about enzymatic catalysts is that they enable and accelerate almost all reactions in the body by millions or even billions of times! Many reactions in the body—such as the production of proteins within the cells, or of DNA itself—would take hours, days, or even months without the presence of the appropriate enzyme. This is particularly noticeable within the cell's mitochondria, whose main function is the manufacture of the "energy currency" in living things, the molecule ATP (adenosine triphosphate, see box). In order for them to do that, enzymes—which are themselves proteins—carry out their work by somehow (there is still a long way to go to fully understand how exactly they do it) enabling the quantum tunnelling of electrons and protons (see box).

Life via a tunnel

Quantum tunneling is the ability of atomic particles to "go through" an energy barrier as though it didn't exist. It's a direct result of the probabilistic nature of their properties. Since they have almost any energy (in superposition), some of them find themselves on the other side. Real quantum "magic".

Energy calculations show that the Sun can't produce the necessary temperature and pressure needed for hydrogen fusion to take place; and yet it does, because of quantum tunneling.

Enzymes, like all catalysts in nature or used by the chemical industry, enable and hugely accelerate reactions, without themselves taking part in them, also because of quantum tunneling, we think.

As mentioned before, metallic ions are essential for the operation of catalysts. Platinum, palladium, nickel, and vanadium are often encountered in chemical production. In the body, we need metal elements such as chromium, magnesium, molybdenum, iron, cobalt, copper, zinc, selenium, and iodine, many of which are essential for the production and operation of enzymatic proteins.

Quantum tunnelling of sub-atomic particles is an extremely important phenomenon and very widespread in nature.

In the case of vegetables and fruit (and all other living things), most chemical reactions in their cells also would not proceed as they do not have enough energy. But the presence of the appropriate enzymes allows these reactions to proceed almost instantaneously via quantum tunnelling. And it is those vegetable and fruit enzymes—or more correctly, the metal ions in those enzymes—that we take from our foods to be able to produce our own enzymes in our bodies. But the metal atoms required to produce the very wide variety of enzymes we need are found in a range of different vegetables and fruit. This is one of the reasons why it's crucial to eat a wide variety of fruit and vegetables—to obtain all the different metal ions we need. Eating a lot of meat is not really a good substitute for them, as it will only contain the metal ions that the animal ate in the first place.

Irradiation for Long Life

Years ago I remember that all potatoes had many sprouting buds and most of the flesh below these eyes was green and would very soon spoil. Nowadays, all the potatoes we buy—unless directly from a farm—appear nearly perfect, without any buds at all. At the same time, vegetables and fruit at the supermarket appear to last for a very long time without spoiling, even if they can obviously become overripe. So what is going on?

What we are witnessing is the result of irradiation by gamma rays or X-rays,[7] a method of preservation which has extended the shelf life of many vegetables and fruit, saving huge amounts of food from being thrown away and protecting our health at the same time.

Let's see exactly what irradiation of potatoes (and nuts, dried fruit, onions, etc.) actually does. First of all it stops sprouting of potatoes and other vegetables. As I mentioned, fresh, non-irradiated potatoes stored in the kitchen will very soon start developing buds which are simply the beginning of a new potato plant. The problem is that budding of these sprouts is accompanied by the production of glycoalkaloid toxins (mostly solanine, a natural defence mechanism) which can act on the central nervous system with quite harmful effects. This is always accompanied by the greening of the potato in the vicinity of the bud, almost as a warning. While small amounts of these toxins are not worrisome, it's always a good idea to remove all green regions before cooking. At the same time, sprouting of buds leads to a break-down of cellulose and the potato becomes soft and mushy with a bad smell. Essentially, the sprouting helps to break down the protein polymers around it (producing amino acids) in order to feed the growing plant.

To stop the above reactions, irradiation of potatoes and other vegetables partly dissociates the DNA in the sprout (and the DNA of any bacteria and fungi present) so it cannot bud, avoiding all these problems and extending its shelf life.

[7] Mostly gamma rays from Co-60 or Cs-137 isotopes and X-rays from electron beam irradiation of a copper target.

Other similar applications and benefits of the irradiation of vegetables include mould and insect control in wheat and fruit and vegetables, and the sterilisation of herbs and spices that could be affected by harmful bacteria and moulds. Gamma irradiation is also used in many countries to sterilise red meat and poultry, as it kills most bacteria and parasites. Gamma radiation is situated at the high-energy end of the EM spectrum, in fact with the highest energy. This makes it quite penetrating, so even large amounts of food can be irradiated together.

You might well ask, is gamma radiation safe? The answer is that it is perfectly safe and we should all be glad that this method has been invented. The energy of the gamma rays and X-rays used for vegetable irradiation is not high enough to activate the atomic nuclei of any of these foods, so none of them can ever become radioactive themselves. The only drawback that sometimes occurs is a slight softening if sprouting has already begun.

And since we are talking about irradiation, there is a theory that life itself might have been kick-started—yes, aided—by the Sun's high-energy radiation, some four billion years ago. It's a difficult theory to confirm of course, but it appears that, at the very early stages of the young Earth, gamma irradiation would have caused many mutations, since there was hardly any protective atmosphere at that time. Some of those mutations would have happened to be beneficial, helping life to take hold and develop.

Sugar, aka Brain Food

Enzyme-mediated metabolism is particularly critical for our brain. And the brain needs only one food: glucose. Yes, the same sugar-derived glucose that we have often been warned against is critical for all brain functions. Many pathological brain problems are directly related to poor metabolism of glucose or its low availability.

The energy consumed by the brain is phenomenal. It has been calculated that, even though the brain weighs just about 2% of the human body, it accounts for more than 20% of the total energy consumption of the body, and it is all derived from glucose metabolism. This means that a normal brain (of mass about 1.5 kg) consumes some 5 g of glucose per hour, or 120 g of pure glucose per day. That's a huge amount and most of it is converted from other foods—nobody (well, almost nobody) can eat 120 g of pure glucose per day. Most of this glucose is used to produce energy that enables synaptic connections between nerve cells in the brain. In other words, it allows us to think, talk, smell, remember, create new memories, and all the other normal brain functions![8]

And all these brain operations are enabled by—what else?—quantum mechanics and in particular the quantum tunnelling effect we talked about in the last chapter. In the brain, quantum tunnelling is not only responsible for enzyme-induced metabolism, but it is also the central mechanism by which neurons connect with one another at synapses. And it is those synapses that allow the brain to operate and carry out all its functions, especially memory function and recall. Some synapses are mediated by chemical messengers, while others are mediated by electrical impulses (brain sparks!) between the neurons.

So the glucose derived from the sugar we eat is responsible for me writing this and you reading it. We couldn't do any such thinking without the sugar. And, yes, we humans do sometimes store a little fat because of a bit too much sugar, but without this we wouldn't have reached the high level of technical civilisation we enjoy today. Of course, too much sugar is well known to lead to type 2 diabetes, so don't overdo it.[9]

What about sugar in cooking? Well, you don't have to add sugar directly to your pot to sweeten food. Adding wine to a frying pan converts some of the alcohol to sugar, balancing out some of the intensity and sourness of onions and tomatoes, the basis for a good sauce and a tasty stew. In some cases, adding just a bit of sugar helps to produce thicker sauce, especially if the food contains some oil and you add some lemon later. Surprisingly for an acidic agent, lemon itself will actually moderate the sourness of tomatoes and other vegetables if it is added at high temperature and allowed to blend for a while. What happens is that the lemon helps to convert some of the protein to sugar producing a smooth and almost sweet flavour.

[8] If you want to lose weight, you could do worse than to read voraciously, carry out intricate mathematics, and solve difficult puzzles.

[9] The widespread misunderstanding (or misinformation) that has convinced people to avoid natural fat and butter has forced companies to add too much sugar and salt to ready meals and snacks to improve their taste. Without some type of lipid (fat), all food has little taste.

Cutting Molecules

In preparation for cooking we cut up many ingredients and it's worth considering what actually happens when we do so. Deep down the act of cutting involves separating molecules or even atoms in some cases! Although the edge of a knife is much larger than a molecule, the final state of separation naturally involves breaking and reforming molecular bonds, or, in some cases, even atomic bonds. The problem is that, irrespective of whether it is meat or vegetables, cutting through raw food will expose fresh atoms and molecules.

Centi, milli, micro, nano, pico, femto

Size estimates are very useful in life and in science. A child is about 1 m (metre) tall. Her finger is 100 times smaller, about 1 cm (centimetre) wide. Her nails are 10 times smaller, about 1 mm (millimetre) thick. Her hair is 20 times thinner still, about 50 μm (micrometre) wide, the same size as a typical skin cell. A mitochondrion or a bacterium is 50 times smaller still, about 1 μm, one millionth the size of the child.

A thousand times smaller still is the size of a small folded protein molecule, about 1 nm (1 nanometre = 10 angstrom (Å) across, while the smallest atom, hydrogen, is about 20 times smaller still at about 50 pm (picometre). Finally, a proton in the atom's nucleus is 50 thousand times smaller, about 1 fm (femtometre).

This means that these fresh surface atoms and molecules are now free to react with the atmospheric oxygen or water or whatever is in their environment. The greatest damage is caused by oxygen which has a huge affinity for most materials and reacts (i.e., combines) with them to create oxides. As I mentioned earlier, nearly all metals (even gold, ever so slightly) react with oxygen to form oxides, and the reaction is strongly exothermic. In the case of iron, the result is iron oxide or rust, and the process is very slow. Oxidation of silver or copper is also slow and we call the result tarnishing. And of course, oxidation of nearly all organic materials (food, wood, paper, natural gas, petrol, coal, etc.) at high enough temperatures (or in the presence of a flame) results in combustion, i.e., they burn, sometimes very violently. This is because they release a lot of energy, given that the total energy of an oxide is less than the total energy of the original bonds.

All metals and organics, if cut finely enough (smaller than about 10 μm) react and burn violently in the presence of oxygen. Spontaneous fires of fine metal and organic powders in a very dry environment[10] are an ever present danger. Indeed, all workers and instruments in such factories need to be "earthed" at all times to avoid electrostatic sparks. Very fine organic powders (e.g., fine white flour and icing sugar) can oxidise so quickly and violently that a fire-bomb explosion can result. Fires involving fine metal powders are actually even more dangerous and often cannot be put out at all once they start. There have been many fires in factories where metal and organic powders are produced or used.

Small particles with a large area

Have you ever noticed how a little bit of soot makes a mess over a large area? It's all because it has very large "specific surface area, SSA", measured in m^2/g. Soot is made of nanoparticles with an SSA of nearly 1000 m^2/g!

This is true of all fine particles and if the particle material is inflammable, they can be dangerous. The finer the particle size the easier it is to start a fire because very fine powders have a very large SSA, which means lower "activation energy" for oxidation: they need less energy to start burning. For example, 1 g (a teaspoonful) of very fine flour of grain size about 500 nm has a specific surface area of more than 200 m^2, the area of a cinema, and if it's airborne in very dry air, it catches fire easily. Soot also burns but it's less easy to start. Icing sugar is also dangerous.

Going back to the kitchen, oxidation is ubiquitous in cooking. In fact, it is often problematic and we try to avoid it. Once a vegetable or fruit is cut, its surface molecules react with atmospheric oxygen and become oxidised. These are now a completely different food, often rancid if oxidation is allowed to proceed. This can affect the food's taste and colour, and even its aroma. In rare cases, such changes are appreciated as they can be used to create completely new combinations. For example, finely cut apple combined with mint gives a very interesting aroma.

On the other hand, we sometimes cut some vegetables as finely as possible to enhance the release of aromatic compounds by increasing the total surface area. Herbs are always used as finely divided as possible, and fresh parsley, dill, rosemary, mint, celery, etc., are all chopped up as finely as possible before adding them to the pot near the end of cooking (to avoid dissociation of

[10] If there is sufficient humidity, static electricity that builds up on insulating materials is easily dissipated to ground since tap or rain water is slightly electrically conducting. But in a very dry environment static electricity builds up in the air and on particles and can only be dissipated by arcing across to ground.

the sensitive aromatics). If we do not chop them finely enough, the aromatic compounds remain closed in the cells and cannot participate in the combined effect.

Atoms and electron clouds

We now know of about 115 elements, some only made in the lab, but they are all made mainly of (positively charged) protons, neutral neutrons, and (negatively charged) electrons. The number of protons in the nucleus (1 to 115) determines the type of element, while the number of neutrons in the nucleus is about the same, but can vary a little, giving different "isotopes" of the same element. In nature, only one or two isotopes of each element are stable, but we can produce many unstable ones which decay to a stable one by radioactivity of 3 types: γ radiation, electron emission, or alpha particle (helium nucleus) emission.

Electrons can exist anywhere within certain regions around the nucleus, each with a specific energy. These are the quantum-mechanical electron orbitals, and electrons can only "jump" from one orbital to another by acquiring energy.

One would expect electrons, being negative, to end up crashing into the positive nucleus, but this does not happen because of QM restrictions.

By the way, most metals tend to catalyse to oxidation, i.e., they accelerate the oxidation and other reactions in fruit and vegetables, so it's not a bad idea to invest in a ceramic knife for cutting fruit and vegetables,[11] for ceramics have no such catalytic properties.

Historically, the question of what happens when we repeatedly cut up a material was first asked by Leucippus and his famous student, Demokritos in ancient Greece. They realised that after much cutting we should reach a point where the material will not be amenable to any more cutting. That point they defined as "atomo," which simply means "uncuttable". The rest is history.

We know of course that all atoms are themselves made by the same simple recipe: a nucleus containing protons and neutrons surrounded by a "cloud" of electrons in specific orbitals. There are only about 95 or so types of naturally-occurring atoms, with up to 95 protons each. Atoms of each type are actually identical whether they are part of a steak or a turnip. It's only when they are part of a molecule that they have specific chemical properties that we can perceive with our senses and measure. For example, sodium (Na) atoms make

[11] I am referring to advanced, "fine" ceramic knives made of zirconia (ZrO_2) and/or alumina (Al_2O_3), not the cheap coloured metal (fake "ceramic") ones.

up a grey, soft, buttery metallic solid[12] and chlorine (Cl) atoms make up a poisonous, yellowish gas, but NaCl is our everyday table salt.

Interestingly, when we eat salt, our digestive system breaks it down to Na and Cl atoms which are used to regulate many body functions, but as independent atoms (we call them ions in this case) rather than molecules and do not cause any problems in the body. The same happens with everything we eat. Metabolism (and the mitochondria I mentioned earlier) can only work with ions, not molecules.

Red-hot to white-hot

All metallic resistive "elements" in the toaster or the oven or under the black glass–ceramic stove top glow red-hot as energy is absorbed by electrons and emitted as photons in the red region of the spectrum due to the quantum mechanical light emission at higher energies. The temperature of those metallic elements is about 800 °C.

If we increase the input voltage, we provide more current so we increase their temperature, the emitted photon energy increases as well, and the colour moves towards the high energy side. At about 1000 °C the element will appear yellow, while above about 1400 °C it will appear very bright yellow, almost white. See later as well.

I Am Toast, You Are Toast

Making a piece of toast using a pop-up toaster brings into play quite a few physical laws. First of all we need a way of heating the bread. This is done by passing an electric current through the heating elements on either side of the slices of bread. These elements are flat wires made of a metallic alloy of chromium and nickel. This is rather an unusual material in that it does not oxidise (i.e., rust) easily, even at high temperatures. It also has a high electrical resistance—the flow of electrons is "restricted" because of electrical interference between the moving electrons (their electric fields, actually) and the metallic atoms in the atomic lattice of the resistive materials, essentially transferring much of their kinetic energy to the atomic lattice.[13] This means it becomes red hot very quickly. Heating of a current-carrying resistive wire is actually called Joule heating. These properties make it ideal for furnaces,

[12] It can be cut easily, but explodes and burns fiercely in contact with oxygen in the air due to exothermic oxidation.

[13] The actual mechanism of "interference" with the lattice is quantum mechanical in nature and not fully understood, yet.

cookers, heaters etc. Your oven and cooker most probably have it in the heating hobs (hot plates) and heating elements. But why does such electronic "friction" cause the element to get hot (and also our hands when we rub them together)? Again, it has to do with the vibration of the atoms. As the electrons jostle and rub against the atoms in the wires, they give them some of their energy and the atoms vibrate more and more violently. And as we said earlier, this means that the material has more energy, which appears as heat.

Talking of rust (or lack thereof), most modern steel pots and pans and knives and forks, etc., are stainless (i.e., they do not rust) because they are made of steel (iron and carbon) containing chromium (Cr, generally about 18%) and nickel (Ni, about 8 or 10%). This is the meaning of the stamp 18/8 (or 18/10) on many good quality pots, pans, and other utensils. These two elements (chromium and nickel) are indispensable to protect iron and steel from rusting. As I mentioned many times, oxygen in the atmosphere is crucial for animal life but it is also the single most damaging chemical element. Apart from damaging most metals, it causes butter and oil to go rancid and is a necessary ingredient for a fire.

Now let's get back to the toaster. When you put the slices of bread in and press down the slider, you are pushing against a spring mechanism until a small clip is engaged which holds the slice holders down. The spring is stretched and exerts a restoring, upward force on the holders. As you press the sliding switch (or the digital switch), you also switch on the circuit and allow current to flow through the electric wires to the elements, which heat up quickly. The red-hot elements radiate heat onto the bread, while also heating up the air around them and all the metallic elements too. The bread toasts very quickly because of a number of synergistic phenomena. The radiative heat quickly heats up the surface water, which evaporates. Continued heating raises the temperature of the now dried bread to over 120 °C, encouraging the creation of the chemical acrylamide (slightly toxic in large amounts) from the starch. The heating of the bread is accelerated by the hot air rising, carrying any water vapour away. At the same time, the water from inside the bread diffuses to the surface, where it is evaporated and carried away, drying the bread out very quickly. The net result is a dry, browned bread slice with a slight taste of acrylamide. By the way, it is never a good idea to over-brown your toast. Too much acrylamide is considered dangerous and is better avoided. We'll see this again later when we discuss fried potatoes.

Finally, once the set temperature has been reached, a small bi-metallic strip thermostat[14] opens the electric circuit and the elements stop being heated. Simultaneously, the same or another bi-metallic strip overheats too and snaps open the clip holding the slice holders down, allowing the slices to pop up. Some modern toasters employ an electronic thermometer instead of the bi-metallic strip thermostat, but the popping action is still carried out by the second bi-metallic strip, releasing the clip that holds the spring-loaded slice holders down. We'll have some more to say about this later.

Can You Walk on Water?

What do trees, the blood vessels in our fingertips, water insects, and a garden thermometer have in common? They all rely on water having a skin to get their jobs done!

As discussed before, water is made up of free-moving molecules of H_2O, with the component atoms transiently held together by hydrogen bonds. But the surface of any body of water such as a lake, a glass of water, or a water droplet actually behaves just like a skin. The molecules at the surface of the water have more available bonds, so they all join together more strongly than the molecules inside the water, and tend to stay together. If you don't disturb the surface, they'll remain at the surface forever. In fact, if a lake remains undisturbed, the water skin forms a barrier which air cannot penetrate deeper easily, and over time the water will use up its oxygen and become stagnant, encouraging the growth of algae and killing plants and animals in it.

The strength of the bonds between water molecules at the surface of water is quite high and we call it "surface tension".[15] The surface tension of water is high enough to mean that, when a body of water is gently disturbed, it forms ripples which move away at first, but quickly calm down as the surface becomes flat again, almost as though there is a taut skin covering it. And its strength is high enough for it to be able climb up a clean surface and pull water behind it in very fine tubes (capillaries). This is exactly how trees get their nutrients all the way up to their uppermost leaves, and it is how blood gets all the way to the tips of our fingers inside capillary vessels. In every capillary, the amount of water or blood is so small that it sticks to the

[14] A small, bonded "bi-metallic sandwich strip," which bends to one side because one metal expands more than the other, thereby breaking the circuit. We'll discuss it again later when we look at kettles and other kitchen appliances.

[15] Measured in N/m, i.e., force per unit length of contact. Water has the highest surface tension of all normal liquids (0.07 N/m).

capillary walls. This liquid "skin" is able to crawl up the capillary, slowly but easily, pulling more fluid behind it. It's so effective that some trees are over 100 m tall.

In space, where there is no gravity, free water actually forms spherical shapes as the surface tension tries to minimise the total surface area for a given volume. This is because the skin tries to remain intact, just like a balloon, and the sphere has the smallest surface for any given volume. In other words, the water tries to minimise its internal energy by forming a sphere that is completely enclosed inside a water skin. On Earth, the surface tension of water creates separate drops on most, very clean, flat, and smooth surfaces. Again, this is because the water skin tries to minimise its surface area by pulling the surface molecules tightly together into one or more droplets.

The water's skin may seem weak to us, but to an insect it is a rather strong surface. That's why many insects are able to walk on water and even run on it without any effort. Some small lizards like geckos can also do that, by using small air bubbles between the hairs under their legs, which effectively create little balloons there, keeping them afloat.

A water strider using the surface tension of water

You can clearly reveal the effect of surface tension by doing a little experiment in the kitchen. Take a very clean drinking glass and gently and very slowly fill it with water. You'll notice that it's possible to fill it until the top of the water, viewed from the side, is about 2 mm *above* the edge of the glass! This is possible because the skin's surface tension keeps it there! If you tear the water's skin by touching it, the excess water will quickly pour out. The effect will be even greater (over 3 mm) if you wet the rim with some oil beforehand. The water's surface tension is aided by repulsive forces between oil and water keeping the water away from the rim. To conclude the demonstration, add a small drop of lemon, which will suddenly reduce the surface tension, whereupon the water will run off easily. This is because the acid weakens the surface hydrogen bonds.

The surface tension of water can be increased even further by dissolving various foods in it, but not oil or honey, which have a low surface tension. That's why sauces and other foods will stick to a plate and cannot be removed easily even by rubbing. Stains and dirt all stick to a ceramic or metal plate by surface tension. To clean them, you need to soak the stain and reduce the surface tension of the solution using soap, as we'll discuss later.

Certain other substances reduce the surface tension of water by affecting the bonds between the water molecules, as with the lemon above. For example, if you spill water on a very clean and dry smooth surface (like glass), it will tend to form small islands, which means it does not wet the surface well. This also means that the water will not penetrate into any small cracks in utensils. But if you add a tiny drop of washing-up liquid, it will wet the surface and get inside the smallest nooks and crannies very easily. Washing-up liquid contains surfactants whose main effect is to reduce the strength of the bonds between the surface water molecules, i.e., the surface tension, allowing them to spread around any surface very easily. This reduction in the surface tension of the water is exactly what allows washing-up liquid to remove stains and dirt from surfaces, sometimes with a bit of mechanical help, i.e., by rubbing. At the same time, washing-up liquids contain very small amounts of alkali compounds (sodium hydroxide or calcium hydroxide), which chemically dissolve fats and other organics, aiding the cleaning process. We'll have more to say about these actions later.

The reduction in surface tension of water is of course the same mechanism used by washing powders to remove stains from clothes. In that case, alkalis cannot be used (they would damage natural fibres), so the beneficial effect of surface tension reduction by surfactants is aided by mechanical tumbling, adding more energy. Water wets and dissolves the stain and weakens its bond to the surface of the material, from where it can then be removed.

In both cases, heating the water increases the energy of the molecules and atoms, increasing the diffusion of water between the molecules of the dried stain, softening the lipids, and weakening the bond to the material surface.

Dripping Honey and Stubborn Ketchup

Isn't it fascinating looking at the last bit of ketchup refusing to run down to the neck of the bottle so that we can squeeze it out? Well, it's certainly frustrating! It looks as though it's stuck at the bottom or sides of the bottle, yet if we just slightly shake the bottle it happily moves. If we scoop it out we see that it's as thin as honey, yet honey flows around happily while ketchup does not, especially if we warm up the bottle. What is going on?

It all has to do with their "viscosity" and the way their molecules behave under force. Viscosity measures how easily a liquid deforms. Water has very low viscosity and honey has high viscosity at the same temperature. The viscosity of the same liquid is sometimes different (at the same temperature) if the liquid is under constant load, e.g., gravity (this is called "kinetic" viscosity), or if it's shaken or sheared somehow ("dynamic" viscosity).

Now both honey and ketchup are slightly polymerised[16] liquids with similar dynamic viscosities but very different kinetic viscosities. That's because the bonds holding their molecules together behave differently. In honey, the molecules can easily break apart or slide over each another, while in ketchup, although they slide over one another happily enough, they do not like to break apart. To distinguish this behaviour, honey (like water) is called a Newtonian liquid, which means that its viscosity is independent of any shaking, whereas ketchup is a non-Newtonian fluid called a thixotropic liquid, whose viscosity appears high when sitting or hanging upside down, but decreases as soon as we set it in motion, i.e., shake it.

Interestingly, if honey has crystallised even slightly (almost invisibly), it becomes thixotropic and stops pouring uniformly and easily. Honey is supersaturated with various sugars, especially fructose (and about 10–20% water), so it is a metastable solution and its sugars can spontaneously precipitate out as microcrystals. Hence, crystallisation occurs because naturally existing fructose microcrystals act as nuclei upon which other fructose molecules start accumulating, gradually increasing the crystal size. After some time the crystals grow enough to become visible, reducing the transparency of the honey. Such crystallisation is easy to reverse by warming the honey in hot water for some time or in a microwave oven for a few seconds and shaking the (now thin) honey to re-dissolve and re-distribute the fructose. Be careful in case you incorporate small bubbles as you do this.

Some other liquids in the kitchen also behave thixotropically, such as yogurt, thick gravy (made with flour), mustard, even thick custard. Some

---

[16] Actually, they are both supersaturated solutions of short polymers.

of them change behaviour if sheared for some time. Unbeaten cream is a Newtonian liquid and runs nicely over strawberries, but becomes thixotropic after beating and can be spread over a cake even though it feels almost as soft.[17] On the other hand, most sauces and thin gravy (made without flour) remain approximately Newtonian even after extensive thickening, and drip easily from the spoon.

By the way, raising the temperature decreases the viscosity of Newtonian fluids, but has less effect on the viscosity of non-Newtonian fluids. Clear honey becomes watery by warming it up slightly, but ketchup and crystallised honey do not.

Not all non-Newtonian fluids behave in the same way. You may have noticed that when you mix water and a little flour together (without oil), the consistency remains approximately the same under slow mixing, while beating is difficult. This is called a dilatant mixture. It behaves this way because the water–starch bonds break with difficulty but slide easily, as in honey and quite the opposite of a Newtonian liquid. This is how the famous "silly putty" is made. It is soft and malleable when squeezed slowly but becomes very hard and elastic when hit with a hammer. In the old days, this dilatant mixture was used to starch clothes (when dry it allowed a slight bending of clothes without forming cracks) and as a fairly strong filler glue for repairs.

To avoid this dilatant behaviour, we add some oil or butter into the mixture to make a dough for pies, quiches, and tarts. This is also the basic recipe for making elastic dough for croissants, pizza, and the famous Greek tsoureki, a light cake bread. On the other hand, we can retain some of the dilatant behaviour by using only a very small amount of butter to make crumble pastry.

By the way, clear honey displays some fascinating behaviour when dripping. Take a wooden honey stick (with or without grooves), dip it in honey, lift it off the jar carefully and watch the dribble as you gradually raise the stick. At short distances, it moves in a seemingly random way, but at about 10 cm, it suddenly starts coiling smoothly, forming beautiful coils just like a coiled rope, even though you keep your hand perfectly steady. At greater distances, the coiling becomes tighter until the motion gets complicated again. Clear honey is also one of the very few liquids whose dribble remains unbroken even at very great lengths, several metres in the case of moderately thin honey, whereas water starts breaking up after just 10 cm or so. Both of these

[17] Strictly speaking, ketchup is pseudoplastic-thixotropic and cream is rheopectic-thixotropic, but let's not go into too much detail.

intriguing behaviours are related to the way honey's viscosity changes with time, as well as its complicated fluid flow.[18]

Honey must be one of the most intriguing liquid foods we have, with sweetness as an added bonus. Because of its high sugar content, and other ingredients, it is also able to break down microbial and viral membranes (it does this electrostatically), and the ancients used it extensively for dressing wounds and for its anti-microbial properties. In many places, this is still practiced, mixed with lemon, olive oil, and various herbs (origanum, thyme, etc.).

Edible Fibres—Widespread and Tasty

We've all heard about the importance of dietary fibre in our diets, along with its many health benefits. I personally swear by them and the original Mediterranean diet is replete with legumes (also called pulses) for their high fibre content. It's probably one of the reasons that the Mediterranean diet is considered so healthy. Let's have a look at some of the physics behind these benefits.

There are two types of dietary fibres in vegetables and fruit, categorised by whether they are water-soluble or not. The non-soluble fibres, such as wheat husk and many fibres found in fruit, are essentially hard natural polymers and very resistant to acids, but also to alkalis. They are quite strong and hardly absorb any water, but they form composite masses with processed food and this helps with digestion.

The water-soluble fibres on the other hand dissolve in water and form soft, complex polymer networks containing water. They look like irregularly shaped micro-sponges. The interesting thing is that they are full of water, up to 95% by weight in some cases, with pores all around them, a type of gelatinisation; but the water cannot leave because of the strong forces exerted by its surface tension. Water is completely trapped inside, but it does diffuse out very slowly over time, which is good news for our digestive system.

Legumes in particular contain large amounts of both water-soluble and non-soluble fibres, which offers a double benefit. Before we can cook them, most dried legumes need to be soaked overnight. This does not affect the non-soluble fibres but encourages the gelatinisation of the soluble ones. The two types of fibre act together to maintain the shape of the legume. The next day the legumes will already have swelled to more than twice their original

[18] Honey's viscosity and temperature determine its laminar-turbulent boundary.

size and we can pour out the small amount free water remaining. On adding fresh water and cooking them in a pot, the gels retain their structure but absorb even more water, together with the other nutrients we introduce (oils, tomato, salt, etc.).

In the eastern Mediterranean, chickpeas are sometimes dried after overnight gelling, resulting in a hard but very pleasant dried chickpea snack called "stragali," generally eaten together with dried raisins (sultana grapes). Healthy stuff and a bit of guilty addictive pleasure to boot. Of course, chickpeas are the basis of falafel, another staple of eastern Mediterranean and Middle Eastern cooking.

A Burning Hot Pleasure

There are people that would consider a meal bland and almost inedible if it didn't have the kick and bite of a jalapeno pepper or other pepper containing ample amounts of capsaicin. There's no accounting for tastes of course, but I personally find that too much capsaicin overpowers all other, more subtle tastes and aromas in food, so I use just a small amount and certainly not in every dish.

Capsaicin is nevertheless quite an intriguing molecule. It's one of a few substances that are able to react ("bind") with the same sensory nerve receptors that are used to signal exposure to fire or rough rubbing. If any of these occurs, the receptor allows the passage of ions into the neurons, reducing the natural electric potential across the neural membrane (i.e., the natural polarisation that exists across the membrane). This causes depolarisation and a drop in voltage across the membrane. The change is sensed by the brain exactly as if you had tried to swallow a lighted match! Or as if you had scratched your tongue with sandpaper, or even touched your tongue on something very cold. That's where the burning sensation comes from if you eat too much capsaicin. The brain thinks you have swallowed a boiling hot cup of tea.

On the other hand, a small amount in a soup or casserole depolarises the same neurons to a much lesser extent, which seems to actually enhance the sensations and pleasure you get from the other ingredients in the food. The amount is tricky to estimate and it depends on the food, as too much or continuous exposure to capsaicin leads to temporary desensitisation of the taste neurons.

Interestingly, capsaicin is a non-polar hydrophobic molecule and it does not react with or dissolve in water—that's why drinking water to relieve the burn makes it worse by spreading capsaicin everywhere—but it is soluble in alcohol. So, when I add wine in the pot after a quick fry of onions and garlic (or whatever I start with), I make sure I wait till the alcohol has evaporated before I add any pepper. We'll see some more of that process later when we discuss the processes that take place when stewing or making a casserole.

Chameleon Food

Have you noticed how many foods change their colour when heated? In some cases it occurs even with gentle heating or even at room temperature. It's almost as though these foods are acting like a chameleon, changing colour as we cook or when just sitting on the bench. I'm not referring of course to colour changes due to the food going off, but changes that occur during cooking, or very rapidly at room temperature.

Red meat will become a very unappetising grey in a pot, but become brown under the grill or on hot coals. It will also become brown if we add anything sweet or alcoholic to the pot. Onions and leeks become transparent when fried. Many (but not all) apples, celeriac, and potatoes become brown when cut, but they remain bright if they are rubbed with lemon or washed in orange juice as soon as they are cut. Cheese will remain creamy or yellow but will easily turn brown if overheated, for example under the grill. Chips in hot oil remain white until they have lost their surface water, then turn golden and then brown.

Gaudy Greek statues

While we have all admired ancient classical statues in all their pure and pristine white marble glory, it looks like we may have been wrong all along. Mineral (metallic oxides) pigments have been found on many ancient Greek statues and there is a colours-on-statues rethink going on at present.

From a number of recent studies, it has now emerged that many—if not all—classical statues were actually painted in such a way as to make them as realistic as possible. Bodies were painted pink or ochre, lips red, hair black, etc. Even clothes were multi-coloured. It almost looks like the ancients may have had a Las Vegas-type of aesthetic after all.

There are many reasons for colour changes and nearly all are connected with some chemical reaction such as oxidation (reacting with atmospheric oxygen), caramelisation, which occurs because of over-heating (pyrolysis) of sugars, the Maillard reaction (heating starch and sugars with proteins), and many others.

Most colours in foods are due to organic molecules ("dyes") and the way they absorb specific wavelengths of light and reflect others back. But oxidation of certain metallic elements in foods can produce inorganic pigments. A good example is haemoglobin in blood, which is bright red when the iron it contains is well oxygenated and forms iron trioxide (Fe_2O_3—rust (!) in arteries), but turns dark red when the blood has deposited much of its oxygen and thus converts to black iron oxide (FeO); that's why veins are dark. Cooking meat releases much of the bound oxygen in blood and turns meat grey.

Inorganic pigments are actually very widespread in nature. Because most are oxides of metals, they are very stable and are not affected by the Sun's radiation, whereas all organic coloured dyes quickly break down under the influence of UV radiation. In fruit, vegetables, and some shellfish, we find them brightest on the outside due to oxidation by the oxygen in air. Some ubiquitous examples are citrus fruit, tomatoes, aubergines, crabs, lobsters, and others. They are all so stable in the sun that they will remain brightly coloured for many years. Metal oxide pigments are very widely used in industry too, to colour plastics, paper, ceramics, glazes, etc.

In nearly all cases, the application of temperature will increase the rate of colour change or the colour will change further. And in most cases, enzymatic catalysis will be to blame.

But why does the food actually change colour during such reactions? The answer of course is connected with the shape of each molecule (before and after the reaction) and the way the molecules absorb light waves. As we saw previously in the book, the colour we see is simply the specific mix of electromagnetic waves that remain and are reflected back after the food has absorbed every other wavelength. So, freshly cut apples reflect nearly all the visible wavelengths, which makes them look almost white, but after a few minutes in the air the cut surface oxidises, and looks brown because the oxide absorbs most of the blue (high energy) wavelengths. The same happens with fried chips. Initially, they look creamy but once the surface starch has turned to acrylamide, most of the blue wavelengths are absorbed, allowing only the red, green, and yellow to be reflected back and thus making them look brown. The same happens with all starch-rich foods, like bread, cake, etc. And if you allow the reaction to continue, all the light will gradually be absorbed and the food will turn black (from the carbon in its molecules, see later), reflecting nothing back except infrared heat waves, which are invisible to us.

Green plants by choice

Photosynthesis in plants is mediated by chlorophyll, a molecule that is able to absorb most wavelengths of light except green, which it reflects back. The remainder of the light energy is used to convert carbon dioxide to sugars for growth, releasing some oxygen.

Evolution being so incredibly creative, it is interesting that plants have not evolved the capacity to use the green wavelength of light. Or maybe they tried, but it wasn't so beneficial after all. If they did manage to utilise all wavelengths, their leaves would look black, giving them high emissivity. This means that they would absorb a lot of light energy during the day but also lose too much heat at night, perhaps freezing them easily. Indeed, at high latitudes, tree leaves are actually a very dark green, a probable evolutionary adaptation to absorb as much energy from the weak sunlight as possible. They also contain anti-freeze compounds to keep them safe at night. All evolutionary compromises, as with everything else.

The same principle applies to all colour effects of course. A red apple absorbs all blue, green, and yellow wavelengths and reflects only the red, while a blueberry absorbs all red, yellow, and green wavelengths and reflects only the blue. Boiling both will change the ratio of blue-red absorption, changing the colour.

Now we can understand why most plants' leaves are green. It is because healthy chlorophyll can only use (absorb) the specific energies of red and blue wavelengths (and some yellow) to carry out its work converting CO_2 and water to sugars, but not that of the green wavelength light, which is therefore free to re-emerge.

Fascinatingly, some foods fluoresce, i.e., they emit a different colour light when lit by high frequency light, for example ultraviolet (UV) light (sometimes called "black light"). As we discussed earlier, UV has a higher frequency, higher energy, and shorter wavelength than visible light. When you shine it on some fruit or vegetables (for example, green chlorophyll-containing, leafy salads, olive oil, ripe bananas, and some berries), some of the energy is absorbed by the low energy electrons in the surface atoms of the fruit, thereby "exciting" them. When these electrons then drop back down to their original ("ground") energy state, they emit the energy difference as light of a specific wavelength. It's a quantum mechanical phenomenon, as we discussed earlier. For example, when olive oil is lit by UV-B, it emits a pink-orange colour, while chlorophyll-laden leaves emit a red light. On the other hand, some berries emit a purple colour.

Imagine a fruit salad made with fluorescent fruit under black light. A sight to behold.

Eggshells—Nature's Engineering Marvel

The shell of an egg is a fascinating thing. It is hard enough to protect the egg as it tumbles away after laying, but it's not so strong that a tiny little chick can't break through it when it is ready to do so. However, an eggshell also has a number of interesting characteristics that are equally fascinating from the physics point of view.

The egg shell is actually made of three layers. While the outer layer—the calcium rich shell that we see[19]—is hard and porous, the inner two layers are semi-permeable membranes with some air in-between. This air pocket is what holds the oxygen that the developing embryo uses to breath and the carbon dioxide that it exhales. The pores help to replenish the oxygen from the atmosphere and allow the exit of the carbon dioxide. Evolution has ensured that the semi-permeable membrane is just right for allowing oxygen to diffuse in and carbon dioxide to diffuse out, but not the other way around. This is achieved by having suitable channels in their atomic structure that effectively behave like one-way valves for each of these gases. This is the reason that eggs remain fresh for such a long time, and even longer in the fridge. In fact, even if the hard eggshell is cracked, the egg will still be good for days, as long as the innermost membrane is not damaged.

This semi-permeability of the egg's inner membranes is something that scientists have been trying to develop in plastic membranes for years, without much success. The "cling-foils" or membranes we use in the kitchen are wholly inadequate to protect foods from spoiling over more than a few days, even in the fridge. They are all made of polyethylene (we'll talk about artificial plastics later), which looks impermeable but in fact is not at all, not even to liquids, let alone gases. Try an experiment: put a bit of water in a small plastic bag and hang it up somewhere. In just a few hours the water will have clearly penetrated to the outside. And this for a plastic bag about 30 μm thick, whereas the inner egg membrane is much thinner still. In order to ensure low permeability, gas barrier plastics are made much thicker or consist of many layers. In extreme cases, when impermeability is critical to preserve aromas or moisture, etc., the plastic is coated with aluminium metal on the inside, as is done with various cake wrappings or ground coffee packaging. I have included an introduction to oxygen permeability later on in the book.

The shape of an egg is also designed—by evolution—to reduce mechanical stresses and minimize the chances of cracking. A spherical egg would be ideal for mechanical strength, and many birds do lay spherical eggs when there is

[19] Made of nano-crystalline calcium with a little magnesium carbonate and a few thousand pores.

little chance of the egg rolling away. A sphere is the configuration that ensures the least surface area for a given volume of material, so it saves calcium to lay a spherical egg as well. From a mechanical point of view, you can hardly beat a sphere or something close to a sphere. It is perfectly possible to place a chair on 4 raw eggs (making sure the chair legs are well padded to spread the load on each egg) and sit on it without breaking the eggs. That's why it is so difficult to break an egg against a soft, flat surface.

Egg shells have another interesting property. They are wetted (attracted) by the egg white much more than by pure water. Have you ever noticed that, when you break an egg in a bowl and a small piece of shell falls in, it is almost impossible to fish it out using a spoon or your finger? It always seems to slip away and escape at the last moment. The best way is to use another piece of egg shell. It will wet and penetrate the white and pin the little piece down very easily.

Shelling a boiled egg is sometimes a real challenge. But it becomes easier if you throw the still very hot egg into cold water immediately after boiling. The sudden change in temperature contracts the egg shell and the membranes much less than the boiled white (a solid polymer now), which makes it easier to shell without a mess.

Finally, can you tell the difference between a fresh and not-so-fresh egg? Just throw it in water. Because the egg shell slowly allows air into the little air pocket at one end, a fresh egg will sink straight to the bottom, but an egg a couple of weeks old will stand on its end at the bottom. And if the egg floats, well, better smell it before you eat it.

The albumin of the egg (10% of the white is albumin, the rest is water) is also a good indicator of freshness. As the egg ages, it loses some water through the membrane and the shell by diffusion, and it becomes more alkaline, weakening and loosening some of the protein coils in the yolk and white, as well as in the surrounding membranes. When you break an older egg in the pan, the albumin and the yolk tend to spread out instead of keeping their shape.

Talking of albumin, it's worth noting that eggs are often used as binding agents or as emulsifying catalysts. We'll talk about emulsifiers later, but the albumin is also an exceptional binding agent and glues together many types of meatballs so that they keep their shape during cooking. Eastern Mediterranean cooking offers many kinds of these dishes, such as the Greek kefte (kofte in the original Turkish), which are fried, or the Greek yiovarlakia, which are made as velvety soups with egg-lemon sauce. Without the gluing effect of the egg (don't beat it if you want maximum adhesion), such meatballs would fall apart.

Collagen and Gluten—Super-Elasticity in Nature

Pinch your skin and let go. Does it return to its original position within a second or two? It certainly should and if not, drink some water, you're probably dehydrated.

Skin is an extraordinary material. First of all, it is by far the largest organ in your body (a popular puzzle question). It consists of billions of cells, all connected together and to each other via nature's wonder glue: collagen. In fact, collagen is everywhere in the body, as it forms the scaffolding upon which all organs grow. It makes up about 30% of the total protein in the body.

Collagen is amazingly stable and has excellent thermal and mechanical properties[20] which give skin its stretchability. At the molecular level it actually looks like a triple helical bundle of macromolecules (another polymer made of many types of amino acids, the basic protein groups), connecting and holding cells and other structures together. It is everywhere in the body, from skin to bones to muscles to ligaments and tendons and everywhere else. It is very important to include it in all foods as it is critical for the health of all our tissues, specifically ligaments, joints, skin, and muscles. One of the reasons (and there are many) that processed meats are not as healthy as unprocessed meat is because they contain little of their natural collagen.

Because of its properties, collagen is not easily broken down by chewing and it is slow to digest. Indeed, it's one of the reasons why we have to cook our food. It melts at a relatively low temperature (about 70 °C) but it remains stable as the fibrils retain their structure for a long time. If you want very tender meat, you need to cook it at low heat for a long time.

Fortunately, in pot cooking, many of the collagen short chains remain in the food, so, from this point of view, pot-roasts and stews are best to meet our daily collagen needs. This is not the case for high temperature cooking, such as casseroles and barbeques, where the surface of the food is exposed to very high temperatures, destroying all collagen and most other proteins too.

Skin is nearly three-quarters collagen (if you remove the water), which is the main reason why it is so strong and elastic. Although many people avoid chicken skin or other types of skin (apparently because of its texture—I guess it's a personal choice), it is an excellent natural source of collagen. Personally, I never miss an opportunity for a bit of skin from a chicken or another bird. Very lean pork and beef contains little fat and collagen, and this is one of

[20] Its special properties are apparently due to a "stereoelectronic effect" where the electron orbitals between the atoms overlap synergistically.

the reasons why I avoid it. I find it lacks taste, probably because those two structures are missing.

Unfortunately, collagen itself is not part of the structure of vegetables or fruit. It is only found in meat products. However, because it is made by our body from raw materials (amino acids and various enzymes), it is still possible to produce just enough of it by eating a lot of a variety of vegetables and fruit, especially nuts and legumes (pulses).

Force and its resistance

Forces always come in pairs. Every applied force, anywhere, is always balanced by an equal and opposite resistance to it. If there is acceleration, the resistance is due to the inertial mass being accelerated. If the force is pushing (or pulling) against a body, the resistance is due to the body's atomic bonds. This is equivalent to Newton's 3rd law, which says that for every action there is an equal and opposite reaction. It's the basis of rocket flight.

The same is true for the body. Bones and tendons always work in pairs. Bones are loaded in compression or bending and tendons always in tension, resisting the load. Significantly, sudden movements or impacts or changes in direction can increase the effective load on the bones and tendons up to 10 times, producing dangerous strains. For this reason the positions where tendons are attached to bones are particularly dense with higher strength than elsewhere.

The elbow, knee, and other joints have to act as fulcrums and keep the loads balanced. This results in gradual wear of the joints so the body must continually repair and strengthen them. Ask any basketball or tennis player about their knee or elbow (and wrist) joints. They are always inflamed. They need loads of collagen.

On the other hand, the job of binding and holding vegetable fibres together is carried out by gluten, a protein mix of both soluble (gliadins) and insoluble (glutenins) proteins. Gluten is mainly found in cereals and is the hydrophilic elastic agent that helps to bind cereals to make bread and similar foods. We'll talk more about it later.

Bones and Tendons—Nature's Super-Materials

Finally, let's talk about really strong materials. Have you ever wondered how it's possible that some super-strong men can lift more than five hundred kg[21]? It's the weight of 6 medium sized persons! It's nothing less than incredible. And it's all supported by their bones and tendons.

Bones and tendons are nature's wonder materials. They contain large amounts of collagen, and bones are also made with calcium and sodium carbonate. But the amazing thing about them is their strength.

Dense bones—for example in the spine or the thigh—have a compressive strength of over 1.5 tons per square cm[22]! That means a normal thigh bone can support the weight of a small elephant or a small car! It's truly astonishing. Of course, we only need that huge strength when we jump from a height, or when there is some other sudden increase in loading.

The places on the bones where tendons are attached are the densest and strongest parts. On the other hand, tendons are always loaded in tension and, because they are thinner, they have to be much stronger. The Achilles tendon at the back of our foot or the ones holding our head up have tensile strengths of over 5 tons per cm^2 (500 MPa) and are able to stretch at least a few mm. Not bad at all. Compare this with the strength of structural steel, which is rarely above 3 tons per cm^2.

What about in the kitchen? Well, a large proportion of bones is made of collagen which dissolves in soups and stews. In fact, meat and especially fish soups are an excellent choice if you are determined to help your skin and other organs remain elastic. Some fish bones (especially the head) contain large amounts of collagen, and this makes the soup thick and creamy (add a bit of oil and lemon). As we saw earlier, ideally, you want a slow, low-temperature boiling process to extract the maximum amount of collagen and calcium and even sodium from animal and fish bones, without damaging the short polymer chains. Higher temperatures will readily dissolve the collagen, but will also break it down into its constituent amino acids, so the body will have to do work to recombine them, not always perfectly correctly.

Tendons are so stable that they hardly dissolve or break down. Not much use in cooking. On the other hand, the cartilage in the joints is an excellent source of dietary collagen and it makes an excellent soup.

[21] The current world record holder is Hafthor Bjornsson from Iceland, with a deadlift of 501 kg. Beyond strong.

[22] About 150 million N/m^2 or 150 megapascal (MPa). More than most woods.

Sea Magic

The sea is a physicist's paradise. There are so many physical phenomena you can observe that you lose count. But since we are in the kitchen and we are discussing the cornucopia, let's concentrate on seafood.

By seafood we usually mean shellfish, and there are some fascinating things about them.

First of all, shellfish have developed some of the strongest muscles and tendons in nature. A mussel attaches itself to a rock as soon as it can, using a special polymer glue, and very quickly grows a "beard" (called byssal threads) made of very strong fibres of the same material that strong tendons are made of, namely collagen, but much stronger. Apparently, they can stretch by over 150% and have an ultimate tensile strength of 4–5 times that of a human tendon. That's not all. The main muscle that closes and opens the shell can respond in a fraction of a second and has even greater strength (for its size) than the strongest animal muscle. Other shellfish have similarly amazing powers.

The shell itself is made of calcium carbonate bonded with collagen, but it is coated both inside and outside with a double transparent layer called nacre. This double layer looks iridescent because it tends to trap and partially transmit some of the wavelengths of light. In other words, it is able to split incident light into its constituent wavelengths, like a prism, or as rain drops do to create a rainbow. Most other shellfish have similar iridescent surfaces, producing this very beautiful physical effect.

Convergent evolution

The development of similar, efficient structures in nature is not very rare and happens naturally over time by "convergent evolution".

For example, the very efficient beaks of octopuses, parrots and tortoises are almost identical and optimised for what they are meant to do, even though the three animals evolved completely isolated from each other. Birds' and bats' wings are another example, also from completely different animal families.

Even more intriguing is the independent evolution of the electrical sensory organs used by platypuses, sharks, bees, and echidnas to locate prey.

Finally, echolocation using ultrasound was developed by such diverse animals as bats, whales (including dolphins), and dormice.

Shellfish are molluscs, and so are another group of fascinating creatures: octopuses, cuttlefish, and squid (calamari). Octopuses and squid are similar, and they are all muscle, made from strong collagen with chitin fibres. And as muscles go, they are super strong. Just one arm of a relatively small octopus (about 2–3 kg) has the strength of a human bicep. And they have eight of those arms, each with its own brain, and can grow to 20 times that size. When tightened against a rock, an octopus arm is so strong and hard that it's very difficult to cut even, with a sharp knife.[23] The suckers are particularly interesting. Each sucker works independently, and squeezing against the body the animal wants to grab. By squeezing, it expels water and air from underneath and the sucker is now held in place by atmospheric (and water) pressure. Just like the rubber suckers a glazer uses to lift a large pane of glass or a spy to climb a glass building. Imagine hundreds of these suckers pressed against your arm.

An octopus propels itself in exactly the same way as a rocket does, by expelling water through a special orifice next to its sack. Its mouth is built just like a parrot's (an example of "convergent evolution") and is the only hard part of the animal. It can cut through a crab's chitin armour like a knife through butter.

Seafood includes crustaceans, which are crabs, lobsters, and similar animals, and it is worth mentioning the strength of their exoskeletons, made of chitin, as it is quite remarkable. The main carapace is multilayered and curved, with particularly thick layers at all points of stress, i.e., points where the main pincer arms and legs meet the body. Without going into detail, every part of their body has clear indications of successful evolutionary development over hundreds of millions of years. The pincers themselves are designed, structured, and hinged in such a way as to be able to exert forces of thousands of newtons, even though the chitin itself is a much weaker material.

[23] I should know—I've had quite a few encounters with octopuses while snorkelling.

4

Gourmet Physics

"The test of all knowledge is experiment."

Richard P. Feynman

G. Vekinis, *Physics in the Kitchen*, Copernicus Books,
https://doi.org/10.1007/978-3-031-34407-7_4

There is more than one way to skin a … beetroot. So, there are many ways to cook and everyone relies on various physical principles and rules. The food we get from each method is also quite different. Sometimes, the same ingredients will even give us a different-tasting result because the temperature is different or there is less stirring or a different sequence or some other parameter. Many gourmet dishes started life as a serendipitous mistake!

In this part we'll visit various cooking methods in turn and observe what happens in the pot, the frying pan, and the oven, and even on a barbeque. We'll also consider various apparently "strange" phenomena that occur in the kitchen and look at how successful food storage depends on physical principles.

But first of all, let's see what happens in the pot.

The Physics of Dissolving

We've talked many times about a food dissolving in a liquid, mainly water. If you think like a chemist, then it's obvious, isn't it? One substance dissolves in another, what is there to add?

But what does "dissolving" actually mean? Well, like many other chemical reactions, it is all guided by electrostatic attraction between molecules, based on quantum mechanical principles. And because of that, the most important question is how "polar" the molecule of the solvent is. In fact, probably the strongest polar solvent is water. As we saw earlier its molecule is made up of an oxygen atom bonded to two hydrogen atoms at a very odd angle, 105°. In fact, this molecule is prone to rotation. This gives the water molecule a strong electric field and makes it strongly polar. So, any other molecule in the water will be attracted by one side of the molecule, to be broken up and maybe reformed differently. AS we saw, water can essentially dissolve almost anything. It has been estimated that the sea contains hundreds of minerals and metals which are frequently extracted for industrial use. For example, magnesium, a very important light metal mainly used to make aluminium–magnesium alloys, but also used in animal fodder and in fertilisers, is nowadays mainly extracted from sea water. Even gold is dissolved in sea water, and although it occurs in miniscule amounts per unit volume, the total amount of gold in all the world's oceans is estimated at up to 1.5 million tons!

So let's see how dissolution of a material takes place in water. When a solid material like common salt (NaCl) is mixed with water, its crystal structure (made up of alternating atoms of sodium and chlorine) is immediately stressed by the electrostatic attraction of the polar water molecule, first weakening the crystal bonds and eventually destroying the salt crystals completely, leaving the ions of sodium and chlorine to swim around freely in the water. If we increase the temperature, the atoms in the crystals and the water molecules vibrate more strongly and therefore the "damaging" effect of the water molecules is even more pronounced, so the salt dissolves more quickly. Similar dissolving processes take place for anything we care to put in water, whether it is a solid or a liquid. Some materials take time, but water eventually dissolves nearly everything we can put in it! But, just like water reforms repeatedly, so do all solutes in it.

Interestingly, when salt (or anything else) is dissolved in water, it has the general effect of reducing the overall polarity of the water, so the solubility decreases until a certain point is reached where there are no more polar water molecules left and no more solute can be dissolved.

In the kitchen, the fact that nearly anything we put in a pot of water will, at least partially, dissolve means that the atoms of the various ingredients can come into close contact and form new compounds. That's how we get sauces (emulsions) and it's how gels and soups develop new tastes. The longer we cook something in a pot, the more uniform the result will be, until eventually we get a smooth, thick soup. In actual fact, under most cooking conditions, what we get are large molecules, similar to the macromolecules of plastics. We'll see how that happens in the next chapter.

Boiling Magic

We saw before that boiling is nothing else than the transformation of liquid water to water vapour, i.e., steam. But how exactly does it happen? If you pay close attention to a pot of water on the fire, you'll notice that bubbles containing steam first form on the walls of the pot. These gradually coalesce together until they are large enough for the buoyancy force—equal to the weight of the displaced water, as Archimedes found out 2500 years ago—to push them up. Finally, they release their contents into the atmosphere when they reach the surface.

Actually, there are two other forces acting on these bubbles. There is the pressure of the water around them compressing them (also due to gravity) and there is the surface tension keeping them stuck on the surface of the pot and making them as hemispherical as possible. In fact, in the International Space Station, where there is almost no gravity, there is only one force, namely the surface tension, and the bubbles do not bubble up but continue growing until they all join up to form a single large, almost spherical bubble surrounded by water. This bubble goes on growing until it has devoured all the water around it, at which point it bursts and releases all the steam at once. Cool. But now let's return to Earth and ask: how do the bubbles form in the first place? I mean, how does the water "know" to start bubbling the way it does?

First of all, there are actually two kinds of bubble. On the one hand we have those that form very quickly and dissipate equally quickly when they reach the surface, as they contain previously dissolved air. On the other hand, there are bubbles which contain steam, which form later, when the water is getting really hot. They tend to form randomly and usually at the bottom and the sides of the pot. If we look carefully, they form at points where there are some tiny imperfections. These are called nucleation sites because bubbles tend to form and grow around them. These are the bubbles that concern us here. How do they form?

There is a bit of magic involved: the magic of quantum physics. It all starts with a single molecule of water that acquires enough energy to escape the bonds of its friends and neighbours and decides to fly around as a gas molecule. But in order to do that it has to somehow persuade thousands of trillions of other neighbouring molecules to do the same, almost in unison. In other words, that single molecule gets enough energy from the heat we apply and starts bumping around and gives some of its energy to its immediate neighbours, and those to their own neighbours, and so on. We don't actually know exactly how this happens, but water molecules are quantum particles and exist simultaneously in a range of energy states themselves. As they acquire more and more energy, the higher energy states spontaneously share some of their energy with neighbouring atoms and an almost instantaneous domino effect takes place which creates that bubble nucleus. Before long these trillions of now gaseous water molecules have formed a visible bubble with a distinct surface at which molecules are continuously being added from the surrounding hot water. They all hold together by their surface tension and the rest is history.

Actually, a hot nucleating surface is not absolutely necessary, and bubbles can start forming in the body of the water instead of on the sides or the bottom of the pot.

Now, when a body of fluid (gas or liquid) gets hot, it expands because the atoms and molecules tend to vibrate more (see later as well for more details) and this takes up more space. This means that the density of the fluid is lower and it rises, just as a large balloon rises when the air inside it is heated. This movement of a warmer fluid in cooler fluid surroundings is the convection we discussed earlier and is what mixes the ingredients in a pot when we heat it. It is also the way a convection heater works. Now, by rising, the warm fluid transports energy from lower regions to higher regions, gradually heating those higher regions. But as it does so, it loses energy and eventually falls back down. This is what is observed when water is boiled in a pot, and we'll have a few more things to say about that later.

Drinkable salts

There are many different salts in drinking water, depending on its origin, containing mainly calcium, magnesium, potassium, and sodium as carbonates, nitrates, bicarbonates, chlorides, and sulfates. All these salts are very important for our nutrition and their presence in healthy water keeps our cells ticking over well. For example, sodium and potassium are critical for ensuring cellular homeostasis, where sodium is exchanged for potassium across the cell membrane. Calcium is of course the main ingredient of bones, while magnesium is necessary for muscle operation.

The total content of salts in water is called its "hardness" and is specified by its TDS value (Total Dissolved Solids) as well as its electrical conductivity, both of which are zero for distilled or deionized water.

By the way, some evaporation happens at any temperature—steam molecules can escape even from cold water or even ice (albeit at a very low rate) since there are always some atoms or molecules that vibrate more violently than others and have enough kinetic energy to escape from the cold liquid or solid. As we increase the temperature and ice melts, the molecules mix more easily and have more energy, so there are even more molecules whose atoms vibrate violently enough to enable their escape.

Evaporation at any temperature is easily observed anywhere there is a water spill. After a few minutes on a dry summer's day or even in winter, the spill will have disappeared, leaving behind a tiny amount of "salt" deposit. Such salt deposits are of course clearly evident inside kettles, where water is boiled and evaporated repeatedly. In areas where water is "hard", meaning there are a lot of salts in the water, it's always a good idea to refresh the water in the kettle frequently and not just top it up before boiling it, otherwise you'll just accumulate a lot of salts. By the way, if you want to clean the salt deposits in your kettle (or anything else), just cover the salt deposits with water, add a glass of white vinegar, heat it a little and leave it. Within 30 min it'll be as good as new. Vinegar contains acetic acid, which dissolves most common salts even those generally insoluble in water. Remember to rinse it out well though, or you'll have vinegar-flavoured tea.

Finally, a few words about the effect of salt on boiling of water. It seems that many people believe that adding salt will substantially increase the boiling point of water. Well, it does increase it, but only by a tiny amount, a fraction of a degree for most concentrations used in cooking, and it happens because salt slightly reduces the vapour pressure of the water.[1] But you might have noticed that sprinkling salt in water which is just about to start boiling (say at about 99 °C at sea level) seems to have the opposite effect: the water will start boiling violently. This seems to indicate that the boiling point has decreased! So what's going on?

Well, the reduction in vapour pressure is not the only phenomenon that occurs in salt solutions. There are two counteracting phenomena. Firstly, salt particles act as nucleation centres for the formation of boiling bubbles. The moment they hit the hot water, bubbles form around them almost instantaneously and often quite violently. That's how rain forms too, around dust particles. But as soon as the salt dissolves in the water, the overall polarity of the water decreases (some of the polar molecules have now been "neutralised"), reducing the overall heat capacity (the energy needed to heat the water), making it easier to reach boiling point. That's why the water appears to boil suddenly. Who would have thought that simply boiling water would be so intricate?

High Cooking

You know of course that water boils at 100 °C, right? Well, this is *only* correct at sea level, when the atmosphere is stable and the atmospheric pressure is about 100 kPa (called 1 bar). During a storm, the atmospheric pressure may be substantially below this value, and it turns out that water will boil at a lower temperature! And if the atmospheric pressure increases on a good day, the boiling temperature will also increase slightly, softening foods in the pot more quickly.[2]

At constant pressure, the boiling temperature remains constant, no matter how much energy one inputs into the system. However, if you feed the steam produced into another container, seal it, and heat it, the temperature of that steam can be increased to more than 500 °C, while its pressure will also increase, with enough energy to drive a generator.

[1] This is due to the relative entropy of the gas–liquid phases.

[2] That's the basis for driving steam turbines made to turn generators and produce electricity. By heating the boiling water in a closed vessel, the steam temperature and pressure increase hugely which is then able to turn the steam turbines.

At the boiling point of water, many water molecules get enough energy to escape the liquid. Because of the added energy, they begin to move around very fast and become vapour. This is true for any liquid. The escape energy depends on the surrounding atmospheric pressure. So, if this pressure is lower, as happens up a mountain or during a storm, the energy that molecules need to escape is also lower. For this reason the boiling point of water decreases by about 1° every 300 m you go up a mountain.

This means that if you live at a higher altitude, you'll have to boil food longer or rely on a pressure cooker. Without it, potatoes, peas, legumes, and similar foods will never soften at altitudes higher than about 3000 m (where the boiling point of water is about 90 °C), no matter how long you boil them. On the last camp before the top of Everest, at about 8000 m altitude, where the atmospheric pressure is only about 35% of sea level, water boils at about 70 °C, which means you can't even make a hot cup of soup! Even an egg will never come out hard at that temperature.[3] It won't help to increase the energy input to the pot. All you'll do is waste energy, as the boiling will proceed at a constant temperature until all the water has disappeared, whereupon the solids left over will start heating up, probably dissociating and eventually burning if organic. Agreed, you'll have other things to worry about up there, but the laws of physics are unrelenting and uncompromising.

However, there is a way out. By using a closed pot with a tight lid, you can increase the pressure in the pot a little while heating it. This increases the boiling temperature until it becomes possible to boil rice and anything else. The reason is that a larger amount of energy (heat plus pressure) is delivered to the food, breaking down fibres and collagen more quickly. However, you will pay for this by the inconvenience and a slightly reduced taste, as we'll see later.

Most of us are used to atmospheric pressure not too far from sea level, while many bodily functions work differently at significantly different atmospheric pressures. For example, our tastebuds (and all other sensations) have a slightly reduced sensitivity (and electrical and other impulses to the brain are weakened) when we fly in an airplane. This is because the atmospheric pressure in the cabin is kept at about 80% of sea level pressure (equivalent to atmospheric pressure at about 2000 m). The underlying reason is the reduced oxygen in the cabin,[4] reducing the efficiency of various bodily functions. That's why airplane foods need more salt and pepper to make them

[3] The white only starts hardening at about 65 °C and the yolk at about 70 °C.

[4] Although the relative amount of oxygen in the air is the same, about 21%, the total oxygen available to us is about four fifths of this.

palatable. But there is a silver lining too. When I fly, I often ask for tomato juice. I enjoy this, but only when flying. On the ground I find its taste too intense.

Reduced oxygen in the airplane is also the reason why any alcohol we drink has a quicker and stronger effect on us. The reason is once again the mild hypoxia (less oxygen in the blood) at the decreased atmospheric pressure and reduced rate of metabolism of the alcohol, which remains at a higher level in the blood for longer.

By the way, the reduced pressure in an airplane is also the reason why they always avoid offering too many vegetables and fruit. These can produce wind and discomfort as the pressure difference is greater when flying than it is on the ground.

The Shuffle of Atoms and Molecules in a Pot

You've all seen it. If we add a tea bag to hot water and leave it there without moving it in any way, the tea will take some time to infuse. However, it will eventually do so, although the tea will be tepid by then. This happens because the spread of tea molecules in water is a slow process of "diffusion," i.e., spreading around in the water. As we saw earlier, all molecules vibrate due to their heat energy, whether in water or in air, and they also shuffle and move around incessantly, mixing with the other molecules. But the net motion of the molecules is not completely random. The physical laws of diffusion tell us that there is method in their apparent madness. A molecule will slowly diffuse from an area where the concentration of its particular type is higher to an area where the concentration of its type is lower. You've experienced this many times in the kitchen. Unless you have the extractor on, the smell of cooking will waft around the house gradually reaching everywhere, even the smallest nook and cranny. Where there was no smell before, it will soon appear. Unless there is a draft (e.g. by extraction or at least two open windows), gradual diffusion of the molecules will continue until the whole house smells equally strongly of the kitchen smell. In other words, the concentration of the molecules causing the smell will become equal everywhere.

Bumping atoms

Until 1905, people weren't fully convinced of the existence of atoms and molecules. They couldn't observe them or measure them, so they weren't sure they even existed.

But it had already been noticed by Robert Brown in 1827 that tiny specks of dust or pollen suspended in clean water appear to move in apparently random jerky movements. This "Brownian motion" was used by Einstein in 1905—his amazing "miracle year" when he published 4 foundational papers—to explain and prove the existence of randomly shuffling and scuttling atoms in liquids and gases.

The rate of diffusion is of course related to the amount of energy the molecules have. The higher the temperature, the faster the molecules vibrate and shuffle around and the faster the diffusion will be. Perhaps you've noticed this. In summer, when the air is at a higher temperature, the diffusion of smells from the kitchen is noticeably faster.

You can actually see this happening in liquids too. Gently put a tea bag in a light-coloured tea cup containing hot water. Initially, the tea bag will be surrounded by the darker tea molecules (water with dissolved tea substances), but the remainder of the hot water will remain clear. Gradually, in front of your eyes, the tea molecules will disperse and diffuse until all the hot water reaches the same uniform colour. By stirring to give the molecules a bit more kinetic energy, the diffusion will occur even faster. If you try this with lukewarm water, the diffusion rate is much slower, and with cold water, there will be hardly any diffusion. This is actually a nice experiment one can do to illustrate diffusion in schools. Various coloured substances may also be substituted for tea leaves.

The diffusion of molecules is affected by the viscosity (i.e., the thickness) of the liquid. When we cook, it is always important to stir or baste the food frequently, otherwise the various substances will take a long time to mix and blend correctly. When we add salt to a cooking pot, it will dissolve quickly in the water, but the dissolved molecules or atoms will diffuse too slowly unless we stir it frequently. This is especially important for immiscible materials, such as oil or pepper in water. The diffusion of oil in water is extremely slow, so unless you stir the food, the two will remain separated, or at best in an uncomfortable embrace of tiny droplets, as in a weak emulsion.

Talking of stirring, who can guarantee optimal mixing of thick mixtures? One of the trickiest hand-stirring jobs to get right is béchamel sauce. If one continues stirring in the same direction (e.g., clockwise), all that will be achieved is a thick sauce in the middle and a thin one at the edges. Changing direction occasionally helps a little but is still not perfect. A very good way to homogenise the sauce is to stir in one direction off-centre, moving around gradually, while occasionally changing the stirring direction. This way, the "centre" of rotation itself moves around and the sauce will thicken uniformly. Not even using automatic blenders can guarantee such perfect mixing.

Up Close and Personal

Top quality cooking requires quite accurate temperature control to get the best and reproducible results. The hob settings help, but they are completely misleading if the pot or pan does not make perfect (or at least consistent) contact with an electric hob or if the gas flame is not uniform. Even a slight (sub-millimetre) gap between the bottom of the pot and the electric hob is enough to make a setting of 6 give the same result as a setting of 4. And waste a lot of electricity and money into the bargain.

A key reason for this is the very good thermal insulation properties of air. Air consists of gas molecules, four fifths nitrogen and one fifth oxygen, with a few other gases in very small quantities. These shoot around, occasionally bumping into each other and exchanging a little bit of energy. But even a thin layer of air can insulate one metal from the other because the heat energy exchange between molecules bumping into each other is not very efficient at all. In fact, during collisions in fluids, molecules and atoms do not actually touch each other neither do any materials actually "touch" one another. This is due to their natural mutual electrical repulsion and quantum mechanical rules that only allow certain mingling of their electrons. So, when molecules and atoms do bump into each other, it is not like hard billiard balls at all, but more like soft footballs that have lost much of their air. In any case, just like colliding balls, the faster of the two molecules will lose some energy during the collision while the slower one will gain a bit of energy, keeping the average energy approximately constant. So, only part of the collision energy is exchanged in air and they do not carry the heat energy of the hot surface to the other surface very efficiently. This fact is exploited in producing highly insulating structures for houses, cars, and even padded jackets, by filling them with air pockets.

Actually, in the case of an air gap between a hot hob and a pot this is only half the story. The presence of a thin layer of air not only insulates the pot, but actively carries heat away! The reason is that hot air expands and if it is free to escape, it will do so gladly. In this case, it will escape out the sides and rise around the pot, carrying heat away. And when it does, fresh, cooler air will rush in to take its place, effectively cooling the pot.

This fact is so important that it is always well worth paying good money for good pots with flat, heavy bottoms that will not easily distort or warp. Most pots and pans destined for electric hobs have thick bottoms for exactly this reason. The bottoms are actually made of two or more layers, including a copper layer,[5] which reduce the propensity for distortion. At the same time, the hobs themselves must also be completely level and flat to ensure perfect contact with the pot during cooking. Flat-topped stoves with a glass–ceramic surface may have a similar problem, but induction heating stoves avoid it to some extent, as we'll discuss later.

In fact, because you can never ensure perfect contact between the hob and the pot, professional chefs avoid using electric hobs at all. They prefer gas hobs, since the flame will always make contact with the thinner pot and the food will respond very quickly. It is also much easier to control the cooking process, although not everything is plain sailing, as we'll see later.

Talking of flat-bottomed pots and the need to avoid any distortion, it is never a good idea to pour water in a pan with very hot oil after frying. Frying is done at temperatures way above the boiling point of water and the temperature of the frying oil can easily reach 180 °C. So, apart from the mess it will make (water being heavier it will immediately sink under the oil where it will suddenly heat up to its boiling point and explode as the steam tries to escape), the metal will on one side try to shrink while the other side is still very hot. This is called differential thermal shrinkage (or expansion) and creates very large mechanical stresses in the bottom of the pot which will almost certainly result in some distortion and warping.

For the same reason, it is never a good idea to place anything cold on the hot surface of a ceramic flat-surface stove. Those surfaces are made of a glass–ceramic (a type of strengthened, semi-crystalline glass) and, although they have higher thermal conductivity than ordinary glass, they are still susceptible to heat stresses. At the same time they are also brittle (as are all glasses and ceramics), which means that they can easily develop cracks if they are loaded while also under heat stress. By the way, metals are much more resistant to heat stresses than glasses or ceramics because they conduct heat away very quickly and don't allow it to build up and cause stresses. All these aspects will be discussed later in more detail.

[5] Copper has one of the highest thermal conductivities of all metals. It also has a very high electrical conductivity, so it's used for wires.

More Edible Plastics—Jellies, Sauces, Syrups, and Creams

Gelling during cooking is often welcome, especially for thickening sauces, various creams, etc. But what exactly is a gel? It's actually a weak food polymer and forms by a type of "polymerisation" reaction. This is of course closer to chemistry than physics, but the underlying mechanisms are all governed by physical laws, as is everything else.

So, what is polymerisation? It's simply the linking up of many small molecules (from a few hundred in food gels to millions in industrial plastics) to form complicated chains. In the case of foods, most of the short chains are organic molecules or pieces of amino acids taken from protein polymers in the food. Polymerisation is quite a complicated process, but in general, it takes a certain time at cooking temperatures, the time for some of the weaker atomic bonds on small food molecules to break (due to the energy input), allowing them to join back together in new ways. Such small molecules could be fats or sugars or parts of proteins. For this to happen in a reasonable amount of time, we need to have an acid in the food which helps to weaken the atomic bonds—lemon, orange, or other fruit juices, or indeed vinegar and even wine (alcohol) will do the trick very well, and then help to form a bridge between the lipid we use and the water. After some time boiling slowly (to make a sauce for example, as in shortening), the oils and fats and other small molecules in the food will join up with the water molecules to form a thick gelled sauce—the longer you leave it, the thicker it becomes.

In some cases, free water does not allow easy gelling—the reason is not fully understood but it is believed to be because of free radicals, highly reactive OH molecules in the water. Slow cooking first boils off most of the free water, some of which is still bonded to the food molecules, and once it's gone, the food molecules can start joining up. That's why, when you make a sauce, always add water very gradually and in very small amounts at a time. If you add too much at a time, the sauce will consist of water with some small gelled pieces here and there, a type of curdling.

By the way, if you don't have much time and you don't mind a bit of cheating, you can add some flour to the sauce. The fine particles of the flour help to bind the free water and "nucleate" polymerization, which then proceeds faster and give greater volume. Personally, I think this is an acceptable short cut, but the sauce will not taste the same because of the watery flour in it. It's essentially a gravy. But in most restaurants, creamy sauces are generally made like this.

Finally, a few words about jelly, something that so many children (and adults) love. The gelling mechanism is about the same, since the powder in the packet includes a "gelling agent" (essentially some acidic substance such as citric acid—artificial lemon) which speeds up the polymerization of sugar and water-containing juice molecules.

Creamy Emulsion or Curdled Mess?

When mixing ingredients in water, we can end up with one of three things. A solution, where the water has dissolved the solute (for example, salt or sugar in water), a suspension where the solids dissolve very little but are light enough to remain suspended in the liquid (e.g., milk), or nothing at all, where the ingredients remain separate (oil or pepper and water). But if you want a smooth, uniform sauce, in which various liquids (and some solids) are well distributed without separating out, you need to work at it. This is called an emulsion, and when it's done right, it's always the supreme culinary pleasure. It's different from gelling and smoother.

The simplest (thin) emulsion is simply a cold mixture of oil and a solution of an acidic agent in water, separated into tiny lipid particles to encourage weak bonding. A very popular one in my country is made simply by whisking or beating together olive oil and lemon juice with added condiments (salt, pepper, chopped parsley, thyme) and is proposed very often in Greece to accompany grilled fish. Because it is made at room temperature, the ingredients cannot react together to form a thick emulsion, even with very strong beating, and they tend to separate to some extent, so this has to be used quickly. Adding some very finely chopped herbs (e.g., thyme or parsley) helps to maintain the gel by including other molecules which act as additional molecular bridges.

But making a creamy emulsion sauce—or any smooth cream for that matter—is not that simple. We want to make sure that the ingredients and dissolved molecules diffuse and distribute well, but gravity and surface tension often throw a spanner in the works by separating out and sinking some of the heavier liquids and solids. For example, when making a béchamel sauce, it helps to first mix the flour very well with the molten butter (or oil)—together with any condiments—before we add the cold milk, and the latter must be non-skimmed milk. Even then, it is still necessary to whisk the mixture strongly, and steadily and very gradually increase the heat to ensure that the ingredients are broken up into very small particles or droplets which remain continuously in touch with one another until they heat up. This will enable them to react and polymerise, thereby increasing their density. At this point, the strength of the (electrostatic) bonds in the polymer sauce will be higher than the surface tension of the individual ingredients, and the sauce or cream will remain smooth.

The most difficult sauces to get right are the ones that contain egg and an acidic ingredient like tomato or lemon. For instance, Hollandaise sauce is prepared without flour, and the use of egg yolks beaten with lemon juice means that it is much trickier to get it to emulsify correctly when we add the lipid. We have to be careful when adding the butter (or oil) at the end, because egg white will react immediately and solidify under heat, destroying the sauce. Albumin is particularly sensitive to heat, so heating must be carried out very slowly while beating all the time to ensure proper distribution of all the molecules. A mayonnaise sauce is made in a similar way, but using vinegar. Actually, the egg white solidifies after the albumin protein has partly unfolded and refolded in a random way, cross-linking all over the place and trapping the water inside to form a gelatine-like structure. That's why, while raw egg is transparent, cooked egg scatters light and appears white. This trapping capability of re-folded albumin protein makes it very useful for forming various foams with other molecules inside, like meringue.

The difficulties with emulsion sauces don't stop there. An emulsion sauce is not a stable configuration. Rather, it is "metastable" and under certain conditions its constituents will try to separate out, sometimes causing curdling or separate phases. In the case of béchamel, its main liquid constituents—butter or oil and milk—are mutually repulsive, since oil and butter are hydrophobic (from the Greek for "scared of water", the opposite is "hydrophilic"), while milk is mainly water. This is why we use a bit of flour to help bond them. But the flour preferentially absorbs water so, if the sauce is reheated too quickly or some acid is added, it will separate out easily, causing curdling.

Hollandaise and similar sauces are even more metastable and can curdle even more easily, both during preparation and during reheating, since they do not use any bonding agent. The secret is to beat them very fast while increasing the heat gradually so that the oily component splits into tiny droplets which then can bind to the watery component long enough to start polymerising. In Greek cooking, we make a similar egg and lemon emulsion sauce[6] without flour. After beating well and heating very slowly by gradually adding hot broth (cooked with olive oil), we add it to the hot pot with immediate stirring and shaking to polymerise it rapidly and create a

[6] Generally, only the egg yolk is used, but I find using the whole egg gives a smoother texture, as the albumin aids in polymerisation, even if it is trickier.

smooth, very pleasant sauce throughout the pot.[7] It's a really brilliant sauce for any meat or fish soup, or for stuffed vine or cabbage leaves[8] and similar dishes.

Curdled sauces and creams can generally be rescued by adding just a little milk (or lemon or just water), while heating gently and beating continuously to ensure very fine re-dispersion of the oil droplets. I find that gentle heating in the microwave oven helps these sauces not to curdle. Generally, in creamy sauces, it helps to use higher-fat milk or yoghurt, which enhances emulsification.

Pressure or No Pressure?

There is more than one way to put energy into a closed system. What we generally do in the kitchen to cook food is to increase the temperature of the food, thereby increasing molecular vibrations which enhance the reactivity of the ingredients and also help to soften the molecular bonds in collagen and fibres. This works well for nearly all cooking operations. Nutrients dissolve more quickly when we heat up the liquid (usually water), and collagen and fibres soften more readily at higher temperatures. Diffusion of nutrients is also substantially sped up at higher temperatures.

However, heat also has a negative effect on many sensitive nutrients. Many vitamins and other molecules tend to break down (dissociate) at high temperatures, something which we would like to avoid. One possibility is to cook at a lower temperature for longer period. This is what is preferable for game or other tough proteins which must first be softened (partly breaking down fibres and collagen by marinading in vinegar and wine) before we add other ingredients, such as sensitive herbs and spices.

Now, increasing the pressure on the food during cooking by the use of a pressure cooker (or "pressure pot boiler") has the major advantage that food is cooked faster. This brings substantial savings on electricity. But what happens to the food?

Increasing the pressure of a closed system with the same heat input will automatically increase the internal temperature,[9] and also the vapour point and boiling point of water. This means that, difficult to soften fibrous meat

[7] Alternatively, you can remove some of the hot broth and prepare the sauce separately. I don't, as I like the sauce to diffuse into the food for a while before serving. It adds to the pleasure.

[8] Because it's difficult to get the sauce right, many restaurants just offer béchamel instead, which I consider a bit of a cheat.

[9] According to the ideal gas law that we discussed before, for a closed system $PV = cT$, where P is the pressure, V the volume, T the temperature, and c a constant. In a pressure cooker, V is constant,

and vegetables will cook more quickly and with less water (since the steam does not escape). Sauces also tend to polymerise more quickly in a pressure cooker because of the additional energy that is made available to the food. Diffusion is also faster when there is more energy around, and this means that blending of tastes is also accelerated. This is useful for making legume soups since you don't need to soak the legumes (beans, chickpeas, etc.) overnight.[10] It is also very useful for boiling potatoes and corn and other simple operations. Interestingly, certain very stable spices, such as nutmeg, cumin, and others, need this higher temperature to release their active ingredients, so foods made using such spices come out with more intense flavours, and you can use much smaller quantities.

Toxic spicy pleasures

It's worth remembering that most spices (and many herbs) are actually toxic in large amounts and should be used with caution by people with health issues, especially those with an oversensitive central nervous system or impaired liver or kidneys. For example, cinnamon contains cumarin, which can damage the liver and kidneys, while nutmeg contains a strong psychogenic compound (myristicin) with potentially serious complications.

The toxicity of many plants is a protective evolutionary strategy developed in parallel with many plants' distinctive smell and colour. Foraging animals attracted to a smell or colour remember very clearly which plants are nutritious and which ones give them digestive problems, or worse.

On the other hand, many nutrients break down more readily in a pressure cooker because of the higher temperature. And not only nutrients, but many sensitive herbs tend to break down faster, well before the rest of the food is ready. It doesn't matter how low we have the heat setting on the pressure cooker. Even at the lowest setting, pressure will gradually build up due to the accumulation of steam until the internal pressure reaches the set point of the valve, which is usually twice atmospheric pressure. The water then starts boiling. At this pressure, the boiling point of water is higher than 120 °C.

And therein lies the problem with pressure cookers. To benefit from the convenience of speed, all ingredients are added together in a pressure cooker. But foods don't all cook at the same rate and most foods require careful stirring, which is impossible in a pressure cooker. Molecular (and atomic)

so increasing P will increase T. It's the same law that results in a bicycle tire getting hotter when we pump it up.

[10] However, you should always throw away the "first" water (after soaking or initial boiling for 10 min) to reduce various indigestible compounds and avoid excess generation of gases later on.

diffusion rates are very different between meat and vegetables, and fibrous foods take much longer to soften than starch. For this reason it's nearly impossible to boil pasta in a pressure cooker to the correct "al dente" consistency. By the time the pressure has built up, the pasta is already soft and may stick together. This necessitates pre-cooking certain foods before placing them in the pressure cooker, or adding them part-way through, which seems to cancel out much of the convenience. Moreover, you cannot see what is happening inside the cooker, so you cannot take any steps to make corrections to the food. Most complaints about pressure cookers centre on that fact. For example, if you add too little water or too much water at the beginning, there is little you can do to correct for that.

There are other limitations too. Although oil-based sauces polymerise and thicken more quickly in a pressure cooker, it is impossible to make sauces that require intense beating, such as béchamel, hollandaise, or similar. It is also inconvenient for foods that require initial stir frying of onions and garlic, followed by wine or vinegar, before adding other ingredients, such as when making bolognaise and similar sauces and many stews. Finally, many stews and soups require thickening at the very end, which we encourage by adding an acidic component such as lemon to catalyse polymerisation of the lipids. If we add it earlier, the soup thickens before it is ready, which reduces diffusion of ingredients, since larger macromolecules diffuse more slowly. This situation, together with the need to add less water (all steam is recycled in the pressure cooker), explains why a pressure cooker often leaves a thick, half-burnt residue at the bottom, no matter what the food.

In general, pressure cookers are certainly convenient and you can make a satisfactory soup or a simple stew, but they are not something that I would use for making a more intricate dish.

Cooking by ... Radar

Wouldn't it be nice to be able to gently and deeply cook the inside of a thick piece of meat or vegetable (e.g., an aubergine or potato) without burning the outside first? Well, there is. It's the microwave oven,[11] and it is probably one of the most efficient and convenient machines we have for heating food (especially water-rich foods), since little energy is lost in the process. Microwave photons have a wavelength of about 12 cm and just the right amount of energy to be able to penetrate up to about 1–2 cm into food (depending on

[11] Microwave heating was first discovered over 60 years ago, during the development of radar, which also uses microwaves, as we'll discuss later.

its water content) and heat it from within. That's why they are so quick and efficient at heating up moist foods and soups—very little energy is wasted in heating the surroundings. Compare this with ordinary infrared electric ovens, where the air and the metal panels are heated more than the food, and the net efficiency is hardly more than 30%. Microwave ovens can reach efficiencies of more than 70%.

We'll discuss the way microwaves work in more detail later, but let's look first at how they are used for cooking. The way they heat foods is very different from other methods of heating. They do not share their energy by collisions between the IR photons and surface atoms as all other heating methods do, after which we must wait for the slow energy transfer inwards. They do so by direct "excitation" of the dipole water molecules. In the same way as you can use a magnet to jiggle the needle of a compass, so microwaves can jiggle water molecules billions of times a second, precisely because they are polar, with a positive end and a negative end. Since molecules also vibrate and move around randomly, the jiggling of the molecules increases the vibrations of other polar molecules around them and this adds energy to the system. It's as simple as that, and it works with all kinds of polar molecules, but not with non-polar molecules, since they have no positive and negative ends to be jiggled by the microwave radiation. You can experiment yourself to find out which foods "absorb" energy from microwaves and which don't. For example, many saturated lipids, such as butter and fat are slightly polar and are readily heated by microwaves.

But what happens to food when it is heated (irradiated[12]) by microwaves? The first thing to remember is that microwave photons do not have enough energy to make food crispy or produce a tasty crust on meat or vegetables.[13] All they can do is heat up water, butter, and some oils. So very dry foods cannot be heated in a microwave oven, no matter how long you are prepared to wait. They also have too little energy to polymerise oils to produce sauces. And certainly, they have far too little energy to "damage" any nutrients, as some claim. Considering their very low energy, it is quite surprising to hear concerns about the supposed danger of minute amounts (less than 2 mW, a millionth of the power of a MW oven) "leaking" from mobile telephones.

[12] I guess it's the word "irradiated" that may have given MWs a bad name. People somehow associate irradiation with X-rays or γ-rays and forget that we are all irradiated all the time by visible photons with much higher energies than MW.

[13] As we saw earlier, microwave photons have thousands of times less energy than infrared (heating) photons and millions of times less energy than visible light.

In any case, microwaves are particularly useful for uniform and very rapid reheating of water-containing foods, especially leftovers, which would otherwise dry up or burn in an ordinary oven, while incurring the usual waste of energy. It is very easy to heat milk or melt chocolate or butter. One can also cook soup fairly quickly, but the difficulty in stirring the food inside the oven makes it a bit cumbersome.

Cooking meat and vegetables with microwaves is also very efficient, although the final colour is a highly unappetising grey since they cannot be crisped and cannot produce the pleasant golden brown colour from the Maillard reaction. This can be accomplished separately under a hot grill or on a hot plate for a few seconds after microwave cooking. This results in a very tender core to the exact level requested—rare, medium rare, or whatever—but with a nicely browned surface, and all this without drying it out. In many busy restaurants (even haute cuisine), the meat or vegetables are generally cooked separately and then smothered by the sauce, so any greyness is not much of a problem (don't look under the sauce).

For the same reason, microwave heating is also very useful for adjusting the core of oven-baked roasts or grills, when the surface of the meat or vegetable or fish has browned sufficiently but the inside is still red or uncooked. This is particularly useful for cooking thick and fibrous vegetables, such as aubergines and potatoes, while maintaining their shape and nutrients, something that cannot be done properly by boiling in water.

Wine, Vinegar, and Lemon—A Dashing Trio

I have already mentioned that the preparation of a smooth sauce without flour requires the addition during boiling of an acidic agent as a catalyst for the polymerisation reaction. Dry wine does the job too, as it also contains various bridging molecules, giving a thinner, but quite smooth sauce, as long as it is added at the initial "light frying" stage. For example, when we prepare seafood (shells, octopus, prawns, calamari, etc.) by briefly frying in oil, as soon as the water has mostly evaporated and the seafood is sizzling slightly, we can pour a small cup of white wine on top of the very hot oil. This immediately reacts with both the sea food and the hot oil to produce a thin, smooth sauce. At the same time, the alcohol evaporates and the remaining ingredients of the wine (sugars and aromatic substances) will diffuse into the seafood, giving it that special taste and even texture. This must all be allowed to go on for at least five minutes (longer is better, but do not allow any more frying). However, it must be completed before we add chopped tomato, peppers, or other vegetables or herbs. By the way, if you add the wine while there is still water in the pot, it will not react and the food will have a wine and alcohol flavour which can be a bit annoying.

Stronger alcoholic drinks may also be used, and in this case they can be added at any time. They have a quite different effect because the alcohol has time to diffuse into the seafood before it evaporates completely, while at the same time inducing polymerisation to thicken the sauce. If you combine this with starch, the effect is quite striking. When I make prawns (or calamari, or cockles, etc.) with pasta, I add a substantial amount of ouzo or raki[14] during early boiling of the seafood, wait for a bit for diffusion, and then add the pasta. The dill and salt are added just before the end. The starch dissolved into the water thickens the sauce very well,[15] so what we have here is a combination of polymerisation and gelatinisation.[16] By the way, pasta should only be added when the water is boiling to avoid prematurely dissolving the starch and the gluten. Because it does not contain gluten, rice can be added with the water at the beginning.

Vinegar (containing acetic acid) and lemon (citric acid) are both used as polymerisation agents, but they also have other effects on ingredients. Being acids, they soften and even partially break down fibres in various vegetables, making them more digestible. For example, all legumes (peas, beans, chickpeas, lentils, fava, etc.) as well as leeks, carrots, celery, celeriac, and other fibrous vegetables can benefit from the addition of vinegar or lemon during the early stages of boiling as this softens them and allows the acid to diffuse within them. Olive oil is added afterwards together with more lemon to thicken the sauce. Aubergines can also be cooked in a stew this way, as long as the cooking is at a low temperature. For example, in a pork stew, the meat must first be pre-cooked (with onions, carrots, and garlic) to partly dissociate the proteins, then fried lightly in its own fat, and only then do you add the aubergines with vinegar or lemon and continue to cook slowly till soft. Sensitive herbs are always added last (to avoid oxidation and dissociation of aromatic molecules), except bay leaves and allspice, which go in earlier to enable sufficient diffusion. If you want a thicker sauce, add some potatoes or celeriac before the aubergines to gelatinise the sauce.

---

[14] Both are quite strong aperitifs from the eastern Mediterranean, with 40% or more alcohol. They are both made by double distillation of grapes with skin and stalks, but while ouzo has a strong aniseed aroma and flavor, raki is more fruity. Other similar drinks are pastis, raku, tsipouro, and arak.

[15] For the cooks out there, I also add chopped onion, garlic, and tomatoes too, plus salt and pepper with the ouzo, and boil at least 10 min before adding the pasta.

[16] Restaurants do something similar, but instead of pasta, they add starch (or water saved from boiled pasta) so they have a ready sauce to add to any type of pasta or rice or other carbohydrate.

Vinegar, wine, or lemon juice are also used to marinate and tenderise game meat (but also any other meat, fish, or vegetables) for the same reason: diffusion into the cells and gradual breakdown of collagen and fibres while imbuing the flesh with aromas and flavours. The longer the marinade, the more tender and more aromatic the result, even at low temperatures. In this case, the cooking time can be reduced and the meat eaten rarer than normal as the marinade sterilises the meat as well. The remaining liquids can of course be further cooked with olive oil and perhaps tomato and herbs, and made into a smooth sauce with additional lemon.

Salt and Sugar—A Love–Hate Relationship

Salt (sodium chloride, NaCl) is one of our most critical nutrients. In fact, the sodium ion is one of the most critical elements for life, as it is crucial for regulating blood pressure and heart and muscle function, as well as a healthy nervous system and a balanced immune system. We require a minimum of about 2–4 g of salt per day (about 1–2 teaspoons) for the needs of our body, depending on the amount we lose in sweat and urine (both diffusion-controlled). So don't scrimp on added salt, unless there is a medical reason or you regularly eat ready-made meals or processed foods, which generally contain too much salt.

Salt has many functions in cooking and it has different effects depending on timing, i.e., exactly when it is added to the food. Let's first look at the properties of salt that are independent of other agents.

Salt (and sugar) is a strongly hydrophilic molecule so it desiccates all foods. It sucks out water from meat by diffusion and osmosis (diffusion across a membrane till the concentration of salt is the same on both sides), so it tends to make meat drier and tougher. It's always better to add salt after braising or barbequing meat or fish, rather than before. It does the same with vegetables, but in this case we do need to remove some of the water, so it helps to add salt at the start. When you start a casserole or a stew by frying onion and garlic, always add a pinch of salt to draw out the water earlier and slightly caramelise the vegetables more quickly, without burning them. The remaining salt is added much later, with the vegetables. Potatoes and celeriac are special cases, as they have a strong affinity for salt and will even preferentially absorb salt from the sauce. If you ever discover you've added too much salt in your soup or stew, just add a potato for a few minutes, stirring frequently. It will suck out all the excess salt, saving your reputation (but not if you serve the potato).

Because salt is such a strong desiccating agent, it can be used very successfully for keeping chips and other fried foods crispy for a little longer. Right after you fry something, sprinkle a bit of salt on it. It will absorb all excess water from the surface of the fried food. Because all forms of life (including microbes and fungi) depend on water, salt is also an excellent preservative, as generations of salted cod fishermen and connoisseurs will certify. We'll talk more about this again later.

Taste sensors

There seem to be (more or less) specific tastebuds for each of the five tastes: sweet, sour, bitter, salty, and umami. Each of them is connected to the brain via a different neuron, but the brain often processes their signals in tandem with signals from the olfactory centre.

The taste buds sense a particular molecule by binding to it (transiently reacting with it) and generating a small voltage signal "signature" which is passed along to the brain along the corresponding neuron. In the brain's processing region (the "thalamus"), the signal is then compared with a previously stored signature to determine the strength and quality of the substance.

Now, salt has a strange effect on our tastebuds. It is not fully understood why, but a tiny amount of salt enhances both the sweet taste of sugar and the taste of umami. In other words, using a tiny bit of salt in chocolate makes it taste sweeter. It may have to do with its desiccating ability again, because somehow the electric signals from our tastebuds are strengthened and arrive enhanced at our taste centre in the brain. I hypothesise that this happens because the non-electrically conductive water is reduced at the point where our tastebud molecules encounter a sweet molecule, removing a point of electrical resistance. This only works when the amount of added salt is tiny, otherwise the sense of saltiness will overpower everything else. Ready meals companies know that, and when they prepare fat-free meals, they add quite a lot of sugar, umami, and salt just to give them some taste. You see, fat-free meat is practically tasteless and without that combination of additives it would be inedible. The misconceived trend for ultra-low-fat foods has played into the hands of ready-food manufacturers. Fat spoils more easily (it oxidises readily at room temperature but is also the preferred growing place for bacteria), so by removing fat and adding salt and sugar, they can increase the shelf life of processed foods and ready meals while reducing the possibility of food poisoning.[17] At the same time, you feel hungry more quickly since sugar has far less energy content than fat and it is also digested and metabolised much more easily.

Salt is also very important for improving the colour of fresh bread. It does that by first increasing caramelisation due to the oxidation of natural sugars in the starch, probably by desiccating the starch. Secondly, it slightly reduces the (oxygen-free) fermentation reaction of the yeasts. Both of these effects produce a nice brown crust more quickly.

Salt is also added to most cheeses and yoghurt as a means of controlling the fermentation reaction by their corresponding moulds. And it is added to processed meats (ham, sausages and the like) to make them look pink and more palatable. In reality, processed meats are a horrible grey colour so the addition of (unfortunately) relatively large amounts of salt increases the conversion of nitrites to nitric oxide (because of the free Cl ions from the dissociation of NaCl), thereby changing the colour.

Ideally, salt should be from the sea or at least iodised, as we need iodine for the correct operation of our thyroid gland. This is probably an evolutionary leftover from the primordial times when our ancestors lived in the sea. Sodium is also quite a critical trace element in catalytic enzymes and we should always include it in our diet.

Both salt and sugar have very good antibacterial properties, at sufficiently high concentrations. No bacterial growth or fungal growth is possible in salted fish, nor in properly made jams (provided that all free water is eliminated during boiling). The main reasons are their desiccating effect on the microbes due to osmotic pressure—without water all bacteria die—and the deadly effect of free Cl on the cell membranes and on bacterial DNA, but not in our blood where it serves important functions.

Looking Through the ... Syrup

Sugar has had a bad press for a long time but for a dubious reason. As I mentioned above, because of the need to add taste to otherwise unpalatable or even bitter fizzy drinks, ready meals, and processed foods, manufacturers tend to add relatively large amounts of sugar (together with salt), leading to over-consumption and serious problems of metabolism.

[17] That's not all. Since sugar is no match for the high energy content of fats and other lipids, it cannot sustain us till our next main meal.

However, sugar in moderation is extremely important in nutrition (and the brain) and it is certainly needed in cooking. In certain stews and other pot dishes using tomatoes, sugar helps to balance out the sourness of fresh tomatoes, giving a very agreeable sweet-and-sour taste. It is important to allow the sauce to boil for a while so that the sugar and water can react with the acidic agent in the tomato to give a smooth tomato sauce.

Sugar also serves to stabilise and lighten the batter in cake mixtures before baking, so that the ensuing carbon dioxide has a chance to diffuse through the structure, something that we'll discuss later.

Mediterranean baking of sweets and cakes such as baklava, kadaifi, revani, samali, orange cake, spoon sweets, etc., often involves making thick syrups, and here sugar comes into its own, at least the unprocessed, brown sugar everyone should use. A syrup is another light polymer, made by reacting sugar with water in the presence of an acidic catalyst like lemon, orange, alcohol, etc. By boiling the mixture for 5 min or so, sugar (sucrose) splits into glucose and fructose, which bind with water molecules in the presence of the catalyst. This type of syrup is called inverted[18] sugar syrup and it is particularly thick if made with brown sugar which brings in additional molecules.

When making this syrup, an interesting optical phenomenon occurs. As soon as the sugar has been added at relatively low temperature, the solution becomes semi-opaque (cloudy) because the sugar splits into glucose and fructose before bonding with the water molecules. While this is happening, light is being dispersed throughout the liquid, giving a cloudy, bright appearance. This continues for a while as the solution is heated, but as soon as bonding with the water molecules is complete, the solution very suddenly becomes transparent. If you don't add lemon or another acidic agent, the phenomenon takes longer, but the transition from semi-opaque to transparent is equally sudden.

Sugar's antibacterial and anti-fungal properties have been known for thousands of years. As I mentioned in the previous chapter, the sterilisation mechanism is mainly due its desiccant effect, but it's also due to its ability to damage the cell membranes of microbes and viruses by forming strong bonds with membrane proteins and lipids. It is the same damage that wreaks havoc on healthy cells in the bodies of people with uncontrolled blood sugar levels.

Honey also displays desiccant osmotic capability, as well as being quite acidic. In ancient times (definitely in Greece and China but elsewhere too), deep wounds were treated by mixing honey (rich in glucose and fructose)

[18] It has the opposite optical polarization rotation to the original sucrose.

with olive oil and lemon (to enable bonding), along with various herbs (thyme, origanum, and others), and leaving it on the wound for some time. All of the ingredients work by multiple routes and all are able to destroy bacterial membrane proteins by electrostatic interactions. Apparently, there is even a substance in honey that affects communication between bacteria so that they cannot develop resistance to it. Some honeys also contain hydrogen peroxide, another potent antibacterial agent with highly reactive oxygen ions.

Denatured Scum Always Rises to the Top

Do you like soup? The making of soup is probably the oldest form of proper cooking (I discount simply placing meat on a fire) and it's no wonder it has evolved to yield literally hundreds, globally probably thousands of different types. And yet, the basic principles are still the same. All soups are made by slowly boiling some protein food with a lipid (I always use olive oil and sometimes mix in some fresh butter[19]), adding vegetables and herbs. The end result is a broth which can be converted to a creamy or veloute soup, or, after clarifying, into a clear consommé soup which may be slowly reduced. The various phenomena present are interesting and incorporate some enlightening chemistry and physics.

Proteins are all rather complex polymers made of thousands of amino acids (basic protein units), coiled into very specific shapes. There are hundreds of thousands of different protein types in foods, all made by strict instructions in the food's genome (the collection of genes in the DNA), but that's another story. During heating in water the protein molecules get to vibrate strongly and eventually the hydrogen bonds holding the protein together weaken, the structure partly breaks down, and, by releasing stored energy, it uncoils permanently into a loose polymeric structure, sometimes with fresh cross-linking of the remaining short-chained amino acids.[20] This is called denaturing (it's mostly irreversible) and any acidic or alkaline agent or an alcohol can speed up and intensify the process by weakening the bonds. However, the amino acids remain mostly intact and therefore the total nutritional value of the food is not affected. Heating an egg demonstrates this process very well. Throwing an egg in boiling water immediately denatures it

[19] Never margarine—it spoils the taste and I definitely do not believe it is as healthy an alternative as so much marketing would have us believe.

[20] Cross-linking is a random, complicated form of chemical bonding between different macromolecules, creating a rigid structure. Most hard polymers and epoxies are cross-linked, as are body cartilage and similar structures.

and the egg white becomes solid. However, if we first beat the egg strongly using a whisk or a fork with a bit of water, the proteins in the egg bond temporarily to the water and then we can add it to a hot soup without it solidifying.[21]

The same happens to meat when you boil it for a soup or a stew. The proteins are quickly denatured and re-form by cross-linking in a more-or-less random manner, making the meat hard. During that early period, some of the surface proteins rise to the surface and form an ugly-looking "scum" which, however, is simply an insoluble mixture of loose amino acids with some fat. If you are making a stew or a broth, you can just let it mix in and it will soon be re-absorbed in the soup, although it'll probably remain in the form of small particles. If, however, you want a clear soup (e.g., a consommé), then it's probably a good idea to remove it right away as it will affect the clarity of the soup. Unlike a broth or a stock, a consommé must be completely clear. You can also filter it to remove particles, before reducing it by extensive boiling.

Further boiling the soup will gradually dissolve some of collagen allowing the meat to soften again, at which point you should stop the process as any further cooking will result in severe cross-linking of the proteins and very hard meat. By the way, cross-linking is the reason why overcooked meat appears to shrink as it hardens, even if it's still in the soup. And it's also the reason why overcooked steaks on the barbeque or in the frying pan shrink and become tough.

Denaturing of proteins also occurs when meat and vegetables are slowly marinated in wine with added vinegar. This also dissolves any collagen. After marinating, cooking should be carried out at lower temperature and for shorter time to avoid shrinking and hardening of the proteins.

I mentioned clarifying a consommé and it is worth looking at the process a bit more closely. In order to make a very clear soup (meat, fish, or vegetable), we need to make sure that all protein particles are removed, including those from the denatured proteins in the scum. We can do that by filtering of course, but a lot of the nutrients will probably be absorbed in the filter paper and I'm not sure whether the paper itself might not affect the soup.[22] The correct way to do this is by cutting all the solid starting ingredients into tiny particles—everything including meat, vegetables, herbs, the lot—and then adding them to cold water. Boiling should be done over a low heat for a few

[21] An egg denatures every time we heat it. A very simple method I use for cooking an egg is in the microwave. I use a round shallow dish (a glass "petri" dish from the lab is perfect). Without oil, it will cook in under 1 min (at half power) to a perfect consistency. Go on, try it. You'll never go back to frying.

[22] Even the clearest of consommés will contain many particles at the nanometer size (a billionth of a metre), and these can easily stick to the filter paper.

hours, with very gentle stirring. You will notice that very soon all the finely cut ingredients float to the surface, because they still contain some air pockets, and they will stay there because of the surface tension of water. Each particle is very light, and as most foods are slightly hydrophobic because of their lipid content, they float around, behaving just like those insects that walk on water,[23] as we saw earlier. If you continue boiling softly (with very gentle stirring), the collagen and other nutrients will very gradually dissolve in the water, which remains clear since all solid particles remain at the surface.[24] After a few hours (!) of slow boiling, everything that can dissolve will have and you can then remove and discard the remaining husks, leaving a very clear but quite dense (and extremely tasty) consommé. Salt and pepper corns may be added at the beginning, but ground pepper and other condiments should be added on the plate as they will cloud the clear soup. No oils are used in this soup and the meat must be very lean[25] as oils and fats reduce the surface tension of water[26] and the finely cut ingredients will simply sink after a while.

At the opposite extreme, as we discussed earlier, to make a creamy or velvety soup, you need to add oil or butter (or use non-lean protein) and use as much of the collagen and protein and fat as you can. All of this will help to thicken the soup, together with emulsifiers such as lemon and egg. In that case, it is perfectly fine to use the scum and I never bother removing it, but take pleasure in re-incorporating it back into the soup. In fact, fish soups are an excellent source of edible collagen which is concentrated in the scum, especially the skin, heads and bones of larger fish like cod and tuna.

Dragged Over the Coals or Trial by Fire?

I guess the earliest cooked meal by an ancestor must have been a piece of meat thrown on a fire, or perhaps dropped accidentally on a fire, or simply discovered half-burnt. Funny how many people still do exactly that. It must be the worst way to cook as it destroys the food, unless it is carefully controlled to keep the temperature below a certain limit and avoid all naked flames.

---

[23] They are the "water striders" we saw earlier. A few species have evolved to do that, including long-legged flies and a few types of spiders. And, of course, some fish have also evolved to nab them from below.

[24] In fact, most of the solutes (the nutrients) in such small pieces dissolve faster than if you use larger pieces, because their specific surface area is very large, allowing faster diffusion.

[25] Fish consommé also works well with young cod, scorpion, and similar lean fish. Sea bream, grouper, and sea bass make excellent rich, velvety soups.

[26] They affect the electrostatic repulsive forces of the water molecules.

That is because all proteins can be destroyed very easily at high temperatures and especially in the presence of a flame, and so can fats and carbohydrates (starches and sugars) at higher temperatures. They are all carbon-based fuels and they can burn (oxidise) quite happily as soon as the water is removed.[27] In fact, the proteins start breaking down because of the excess heat quite early on, long before any actual burning. When burning starts (and the surface of the food has converted to primary black coal), it is too late to save it. It is indigestible and in fact it probably contains many aromatic hydrocarbons, all quite toxic.

The only way you should broil food on a barbeque is using charcoal, never directly on wood flames and wait until the wood has converted to coal,[28] and only when all flames have completely subsided.[29] The temperature at the surface of the meat or vegetable should not be higher than about 200 °C to ensure heating without burning. Any flames (e.g. from dripping fat catching fire) must be put out immediately (cold or wet salt works fairly well as it has a high heat capacity and absorbs the heat) to avoid burning the food.

Because of buoyancy (a gravitational effect), hot air rises so a barbeque does not need a lot of hot coals to do a good job, and they certainly do not need to be red hot. Well-prepared hot coals are grey, without any covering of ash, which is an excellent thermal insulator. Ash prevents heating of the air and stops all infrared radiation from reaching the food above, so it should be removed as quickly as it forms.[30] In fact, ash is such an excellent insulator that hot coals in a fireplace retain their heat well enough to be used to start a new fire the next day. Nomadic people cover hot coals with ash, to protect them from further oxidation, and transport them in a skin bag until the end of the trip, when the coals are uncovered and can be used immediately for cooking as they will have retained their heat very well, even days later.

To reduce the rate at which the coals burn, you can reduce their oxidation. While at the beginning you supply a good flow of air to start combustion, once the coals are red hot, you can reduce the air supply to an amount that will give you just the right temperature. This will also reduce the rate of production of ash.

[27] As soon as all water evaporates, food converts to and emits various gases which ignite, before final pyrolysis, which ends with carbon.

[28] Wood fires emit many toxic gases, especially hydrogen cyanide (HCN).

[29] Initially, coal will oxidise only partially, producing carbon monoxide (CO) another toxic gas. Later, it will produce carbon dioxide, assuming there is an ample supply of fresh air.

[30] A vibrating barbeque would do the job very well, by occasionally shaking the ash off the coals. I wonder if anyone has thought of it.

It is all the brain's fault. As we saw earlier, the brain uses glucose almost exclusively for its energy needs, and this can be obtained fairly easily from fat by a simple enzymatic transformation. So, as soon as it detects excess fat in the body (not used up for daily needs), it makes sure it is stored and very well protected until it is needed. Interestingly, the brain converts all excess sugar into fat as well, because it is easier and safer to store fat than the rather toxic glucose. If you want to lose weight, reduce your sugar intake and hold on for a few days, while continuing to use up energy (e.g., exercising and thinking hard). The brain will then be forced to convert stored fat into sugar for its huge needs. But you have to persevere, as the brain will easily go back to its bad (survivalist) behaviour of directing the storage of fat as soon as it can.

The lipids we eat can very easily be damaged beyond reasonable usefulness by cooking the wrong way and by oxidation in a warm environment. Here, a chemical transformation is to blame. Overheating is the main culprit and high-temperature frying the main cooking route to damaged fat. Let's see how these problems occur.

Most everyone loves the taste of fried chips ("French fries") or steaks, but do you know why? It's a drug, of sorts. During deep or shallow frying, starch from potatoes or flour is converted to certain kinds of sugar, which are of course very palatable. But at high temperatures (e.g., during frying) starch is converted to a brown chemical called acrylamide which gives potatoes and bread their lovely golden brown colour. As mentioned before, it is also a bit toxic,[35] so it's a good idea to stop frying when the chips are lightly golden, not browned. The reason why acrylamide is so easy to form by frying is because most oils boil at over 180 °C, while water boils at just 100 °C (at sea level). By heating the oil to about 180–190 °C, the surface water on the food (say potato chips) boils and evaporates violently, the surface of the potatoes is quickly crisped, and the potato is sealed, thereafter reducing oil absorption. Shaking the basket with the potatoes helps to expose new surfaces to the hot oil for an even crisping. During frying, the temperature of the oil–water–potato mixture remains just above 100 °C until all free water has evaporated. At that point, the oil temperature starts increasing fast and the starch starts overheating, gradually converting the now desiccated surface starch to acrylamide. If you wait too long, the acrylamide will get darker and darker, eventually turning black and becoming inedible as more and more of the hydrogen and nitrogen break away and only the core carbon atoms remain. In fact, acrylamide starts forming once the temperature of the oil in the pot has climbed above about 130 °C. This gives us a quick way of telling that it's

[35] Lots of high-temperature transformations in the kitchen give slightly toxic but tasteful by-products. It's a case of balancing taste with safety while our liver expertly eliminates them over time.

time to remove the chips. By the way, chip size matters and they should be up to about 7 mm thick. The problem is that, if the potatoes have been cut too thick, the core will not be properly cooked before too much acrylamide has formed on the surface. If they are cut thinly, they will become very crispy without too much acrylamide—just like crisps in a packet.

While the chips are still frying, it's a good idea to shake them in the basket to loosen them. Acrylamide is quite a hard and adhesive compound and forms tiny bridges between chips (I once observed them under a microscope), bonding them together.

Finally, it is worth noting that most oils also start breaking down (dissociating) above about 160 °C, so frying should only be done using oils that have a higher dissociation temperature, for example sunflower or safflower oil. High quality olive oil should be avoided for shallow frying as it breaks down at much lower temperatures (about 110–120 °C) and in any case it's such a waste to destroy the fruity taste of such olive oil, especially the "extra virgin" type. It should only be used in salads and in pot cooking, where the temperatures never go above 100 °C. By the way, "extra virgin olive oil" is a formal designation to indicate extraction only by the—centuries old—cold-pressing method, without the application of any heat. This is important because olive oil extracted by thermal methods (which give almost twice the yield) is already affected by heat and has already lost much of its fruity aroma and taste, although it can be used for frying.

The air fryer is a recent development which aims at reducing oil absorption, but presents its own problems. The potatoes are placed in a metallic basket and hot air is blown on them at about 160 °C, quickly sealing the potatoes and gradually cooking them internally. The problem is that acrylamide starts forming almost immediately and not at the end as with deep oil frying, while the potatoes still have to be heated for some time to make sure they are cooked properly inside.

Barbequing meat presents another problem, rather more serious. At high enough temperatures, e.g., during grilling on a barbeque with a free flame, fat is converted to various toxic compounds called cyclic aromatic hydrocarbons which are certainly bad for health (proven carcinogens). It's mainly the result of any flame that is produced by dripping fat onto the hot coals, so make sure you put out any fire immediately by sprinkling wet salt on it or a good dose of beer. As I mentioned earlier, salt puts out a fire very effectively, simply by soaking up large amounts of heat and thereby cooling it down. In physical terms we say that it has a high specific heat capacity (energy that needs to be absorbed per kilogram to raise its temperature by one degree) and it is something of a champion in this regard. The value for salt is only about a quarter of that for water, but still very high. Water has the highest specific heat capacity of all normal materials (only hydrogen and helium gases have higher values). This is in fact the basic reason why life arose on our planet, whose average annual temperature (about 15 °C) is ideal for water in its liquid state, providing the perfect conditions for life to get a grip and develop. Water's very high heat capacity and its very high "latent heat of vaporisation" which we discussed earlier means that it can't be heated up and evaporated too easily by the Sun, allowing life to get a firm hold on the early Earth.

Batter Matters

Everyone likes pancakes and honey puffs, I think. And surely everybody likes Wiener schnitzel. I like them both, not only because of their taste, but also because of the fascinating physics underpinning them. In fact, everything that requires coating has interesting features. Let's look at them in more detail.

First of all, let's see how pancakes (or honey puffs or dumplings) are made. We simply make a batter of flour with water (perhaps adding some baking powder or bicarbonate of soda[36] or even some beer) and then throw a spoonful into very hot oil at about 180 °C. If the oil is very shallow (or we use a hotplate at this temperature), we get a pancake but in deep oil we get a batter puff. But what exactly happens?

The flour–water mixture (essentially starch with gluten and perhaps some short fibres if we use whole wheat flour) becomes a thin glue which, on contact with the oil at about 180 °C, immediately converts to acrylamide. The latter is strong and hard, encasing and protecting the soft batter inside, which has since filled up with carbon dioxide and air. At the same time, the batter tries to minimize its surface area (like a balloon), resulting in a disc or spherical shape, and the puffs are ready in a few seconds. The added baking powder forms carbon dioxide bubbles which cannot burst straight away, giving a light, swollen pancake or honey puff. Beer does the same, as it contains (and generates) carbon dioxide bubbles itself, but it's better to wait a few minutes in this case to allow for the batter to swell a little before frying. I have already discussed the effects of using alcohol in cooking.

By the way, flour with a very little water is well known as a glue because of the gluten it contains. In olden times, carpenters and other wood workers used to make their own glue from very fine (gluten-rich) processed (white) flour and water. It had to be used very quickly as it coagulated and

[36] Both contain carbonates which quickly decompose above about 160 °C to give off carbon dioxide.

dried quite fast. But wood stuck with that could remain well glued for many years.

We use this property when we fry fish or schnitzel or calamari. The problem is that fish or calamari are very sensitive and will dehydrate and shrink in seconds in hot oil, so we need to delay this process until the flesh is cooked. For this we employ a thick batter or simply dip and coat the slightly wet fish in flour and allow it to dry slightly to make a strong coating before frying. As soon as the flour starts converting to golden acrylamide, we know that most of the internal water has been eliminated and the food is ready to be served.

In the case of a schnitzel (or a chicken leg for that matter), the added beaten egg and crumbs after the flour give added taste, but they also offer some protection to the sensitive thin slice of meat inside and ensure that it stays moist during shallow or deep frying in oil.

Drink and be Bubbly and Merry

We've already discussed the catalytic effect of a little alcohol on the polymerisation of lipids when making smooth sauces. It does this by being slightly polar. This allows it to bond equally well with water and various lipids, while at the same time breaking down some proteins, which thickens the sauce even further. But alcohol also has a number of other effects on food and food preparation.

As I mentioned above, beer lightens up batter by including carbon dioxide bubbles as it contains various foaming agents. It can even lighten up heavy batters (e.g. bread) because it generates bubbles and at the same time strengthens the bubble surfaces by the presence of a small amount of alcohol, which binds the water with the oil. Allowing the batter to swell a little before baking or frying increases the subsequent swelling substantially.

Fizzy wines (or champagne) also work fairly well too, as the greater amount of alcohol tends to strengthen the additional bubble surfaces and raises the batter better during frying. The additional taste left over by the wine or champagne makes for an extra special treat.

The strength of the water–gluten bonds makes it difficult to create a flaky batter or dough with white flour. A way to achieve this is to use a stronger alcoholic drink—whisky, rum, brandy, vodka, raki, ouzo, etc., all with about 40% alcohol—and stir the batter or knead the dough a bit more. The larger amount of alcohol and the substantial amount of sugar results in weakened water–gluten bonds but stronger oil–water binding, and the batter or dough becomes flaky during frying. In each case, the alcohol is added to the almost ready batter or dough and folded over many times before frying or baking. This is how croissants are made, adding butter as well.

Strong alcoholic drinks are also used for various flambé dishes, for example in Banana Foster or crème brûlée. The alcohol is ignited and the flame envelops the dessert, but the temperature on the dish itself is not very high, so browning by the Millard or caramelisation reactions is minimal. The alcohol binds the dish while evaporating, and once the alcohol has burnt off, all that's left is the aroma and the taste of the drink.

Swelling Pressure and Architectural Perfection

Imagine a world without carbon dioxide (CO_2). Actually, it's impossible, since every bit of life on the planet depends on the availability of this simple molecule: a carbon atom flanked by two oxygen atoms, in addition to water. Plants and algae grow by taking in carbon dioxide from the atmosphere and converting it with the help of sunlight (via amazingly complicated but fascinating endothermic reactions called photosynthesis) to sugars and proteins. Actually, it's another interesting physical process that allows this to happen: the conversion from CO_2 to sugars would take years if it wasn't for the presence of a metal catalyst that accelerates the reaction by millions of times to enable plant growth. Such catalytic acceleration of the CO_2-to-sugar reaction has a number of intermediate stages and for many years it was not understood. In fact, as we discussed before, all life depends on catalysis of biological reactions, enabled by the enzymes in all living bodies. Many of the vitamins and minerals we need for a health life are actually catalysts which enable the necessary processes of life. There is strong evidence that such catalytic processes depend on the quantum mechanical characteristics of the system at the molecular and atomic levels.

But I digress again. We were discussing carbon dioxide. Animal life produces CO_2 continuously by the exothermic reaction between carbon in our food and oxygen in the atmosphere. It's a good thing that plants are able to use this carbon dioxide to produce sugars and other compounds with the help of sunlight. But let's disregard all that for a moment and see where we find it in the kitchen. First of all, we breathe it out all the time and it is also produced by the combustion of the gas in the hob or the oven, as long as the hob is clean and there is sufficient oxygen.

In the kitchen, we depend on the production and controlled release of carbon dioxide for numerous cooking and baking processes. A cake would never rise, and would therefore remain dense, if it wasn't for the pressure inside millions of bubbles of carbon dioxide gas in the moist starch–gluten mixture. This gas is produced by the dissociation of baking soda molecules (sodium bicarbonate, consisting of a sodium atom with a hydrogen, a carbon, and three oxygen atoms) or baking powder, which is a mixture of baking soda with a weak acid like sodium phosphate. As soon as the wet cake mixture reaches about 150 °C in the oven, these will start dissociating and releasing carbon dioxide, creating bubbles in the cake and making it swell and rise.

Actually, much more is involved in making a cake or bread. If we put the cake in a cold oven and let it heat up gradually, it will not rise because the carbon dioxide will gradually seep out at a lower temperature (from about 140 °C) and not do its job properly. It's critical that the cake mixture be put in a pre-heated oven at a high enough temperature (at least about 175 °C), so that the top layer dries out very quickly and forms a barrier to any gas seeping out too quickly. The gas in the bubbles is under pressure from the surface tension of the bubble skin and all the bubbles together push against the surface of the cake. This is exactly how a balloon works: it expands until there is equal pressure inside and outside, taking into account the elasticity of the balloon. Once the cake has risen enough, the temperature of the oven can be reduced slightly (to about 160 °C), so as not to overbake the surface of the cake but still allow the evaporation of the water and the reaction between the starch, sugars, etc., to proceed. Finally, all the water will dry up as it diffuses to the surface and evaporates away, and the cake will retain the puffed-up appearance of the dried bubbles.

Biscuits and cookies are made much the same way, but most of them nowadays, especially the ones referred to as "soft", are first dried and pre-baked in a conveyor microwave oven and then passed under a normal infrared oven to given them a slightly crusty outer surface.

Before we leave cake making, a word about aromatics in cakes and biscuits. Most aromatic compounds are very volatile molecules and are extremely sensitive to baking. They dissociate very easily and their aroma is all but lost if the temperature is too high or baking takes too long. Examples are various "essences" such as vanilla, orange, lemon, etc. In my experience most such essences are lost during baking and we might as well not use them at all.

However, there are a number of ways out of this conundrum. The first is to add the aromatics in a slightly coarse, solid form, and make sure the batter is dense so that the baking time is short and some of the compound remains behind. Examples are small particles of mastic, vanilla, cinnamon, orange, or lemon rind, etc. Alternatively, you could reduce the original amount of sugar in the cake and add the aromatics after baking, in a thin syrup which produces a soft, moist cake of the kind made in many countries of the Eastern Mediterranean or Middle East. Finally, if the aromatic is particularly sensitive, it could be sprinkled on afterwards. An example would be rose or orange "water".

What about bread? Well, the overall mechanism is exactly the same. We need a sticky dough (the more starch and gluten, the stickier it will be) with good surface tension, and we need something to produce carbon dioxide to create bubbles. It's actually quite possible to make bread using baking powder and baking soda, but it won't taste like real bread. Lots of cheap bread products use these and they taste strange or downright awful. For real bread, we need to use yeast, a type of fungus. This is actually a living mould-like colony of millions of spores that produces carbon dioxide by a fermentation reaction, even at quite low temperatures. In fact, real bread is made by allowing the dough to swell at a very mild temperature (about 35–40 °C is sufficient) for about 1–2 h and then immediately baking it in a pre-heated oven at about 180–200 °C. The big difference with cake is that swelling ("raising") must be completed (up to 100% of the starting volume) before baking begins. Indeed, baking does not produce any more carbon dioxide as the yeast is killed at high temperatures. By the way, many porous cheeses such as Swiss emmentaler are produced using the same principle. Various kinds of cheese yeast produce carbon dioxide. This increases the pressure inside the cheese mass, which swells as it tries to equalise the pressure inside and outside, being assisted of course by the strength of the cheese mass around the pores.

Interestingly, 100% whole wheat bread or cake can never be made to swell as much as white bread, or even at all. Because of the additional fibres and reduced starch and gluten, it is impossible to seal the surface and the bubbles completely, and carbon dioxide seeps out well before the bubbles can expand. The best that can be managed is about a 20–30% expansion to make so-called "German bread". For lower contents of whole wheat flour, the sealing of the bubbles is sufficient, so swelling up to 100% is possible again. Interestingly, oat flour is gluten free, but its starch content is high so it forms strong and elastic bubble walls and it can replace up to about 30% of the white flour if preferred, with only a minor effect on the swelling. To get the same swelling, add more yeast.

The science of fracture

There is a whole scientific field called fracture mechanics, which deals with the way that materials deform and fail. The whole field is based on the way materials absorb energy as the atomic lattice deforms, and also on how they dissipate fracture energy in order to sever atomic bonds at the point of fracture. It has been crucial in recognising that thick structures are not necessarily strong, as they can fail by the extension of tiny preexisting flaws. It has also shown that, in many structures, toughness (energy dissipation) is more important than strength in determining reliability.

In the case of composite materials, fracture mechanics explains that hard, blocky inclusions in a "matrix" reduce the strength, but if they are elongated or tough, they can increase the energy needed for a crack to propagate thereby delaying final fracture. Ever since its development 60 years ago, fracture mechanics has been instrumental in the development of large, tough, and lightweight ships, very tall buildings, safer cars, safe pressure vessels, and reliable ships among other structures.

Yeast cannot be mixed with baking powder as some of the sodium carbonate decomposes early in the watery mixture, producing caustic sodium hydroxide (NaOH, a strong base) which kills the yeast.

Bread or cake containing nuts or olives or raisins or anything else that we like to add behaves the same as when adding whole wheat flour as far as rising and final shape are concerned. You can add up to about 10% of such tasty morsels, but you should expect a slightly denser final result because some of the carbon dioxide will diffuse out more easily. This is due to new channels opening up through the mechanical action of these added foods. Adding more yeast or baking powder helps a bit, but not much.

But the biggest effect of adding nuts, raisins, etc., is on the strength and toughness of the final cake or bread. These added ingredients tend to weaken the structure, so expect the cake or bread to come out a bit crumbly, especially if the pieces are too large. Interestingly, adding small currants or raisins to cake has a toughening effect: the currants tend to bridge the cracks and the cake becomes less crumbly. Nut s and larger fruit usually act as "flaws" in the structure, weakening it. On the other hand, oat flakes have a strengthening effect, because of the large amount of starch that acts as additional glue. The strongest and tastiest olive bread should have a lot of whole wheat (I use about 60%), with long, thin slices of olives. The fibres in the whole wheat flour act as millions of tiny bridges across the gluten-rich white flour, giving great strength on top of the fantastic taste and being very healthy. Try it and you'll see.

I sometimes illustrate the difference between strength and toughness to students by using hard chocolates. A normal chocolate breaks apart directly – it is brittle like a ceramic. But the same chocolate with raisins will crack while not separating immediately: it is tougher.

Finally, a quick word about pastry and pies and puddings. All of them are also made of different types of dough, but the mechanism of swelling is approximately the same. Croissants and dumplings rely on repeated layering by folding the dough (a very long process), which encloses hot butter or another lipid and keeps the thin layers separate. During baking, the different layers swell and shift or slide separately, giving the fluffy structure we enjoy. In my part of the world we don't bother with folding but we use many layers of very thin dough (called "fyllo", Greek for "sheet") with which we make pies, or the famous syrup-laden sweets like baklava, etc.

Pasta is of course a type of pastry made by extrusion of very elastic dough through dies, using exactly the same manufacturing extrusion method as is used to make minced meat, plastic cutlery, accessories, and toys as well as aluminium and steel bars and structural bricks with holes. After extrusion, the pasta is dried at low temperature, mainly using microwave driers, to preserve the nutrients. When boiling pasta, which must be done by throwing it in boiling water at high temperature to avoid it becoming soggy and oxidised, water diffuses into the structure, swelling it and softening it again. A pasta pie or lasagne with a topping of béchamel cream[37] or melted cheese is very popular in Greece and Italy, as the pasta is protected from oxidation and from drying by water diffusion while adding a further tasty layer.

Fine Delicacies—Mouldy Yeast and Tasty Microbes

It is a well-known fact nowadays that our health depends on our gut bacteria. Not a day goes by without some new discovery on how our gut "microbiota" controls many of our bodily functions or how the bacteria need some specific nutrients to stay healthy. They control many of our normal digestive functions and they manufacture various hormones and other chemicals important to maintain our health and immune system. They even adjust our behaviour and responses to outside influences. Most hide in our guts where oxygen levels are very low because they are mainly anaerobic (they cannot use

[37] Coarse pasta mixed with a mince and tomato sauce ("Bolognaise"), topped with béchamel and cheese is an extremely popular dish in Greece called "pastichio". I personally prepare it layered with lasagna.

oxygen for energy like mitochondria do for us). The important thing is that these bacteria are critical for breaking down starches, fibres, and potentially toxic molecules from foods to smaller molecules which then diffuse through the intestinal wall into the bloodstream. They also synthesise vitamins B and K and many amino acids using the raw materials we give them. And all they ask to stay healthy is a reasonable amount and variety of fruit and vegetables. An unhealthy microbiota leads to an unhealthy body.

Our friends, our gut bacteria

Our "microbiota" is the flora of our gut (especially in the thick intestine), which consists of trillions of bacteria of many different types, in a close symbiotic relationship with us.

Until recently, our gut bacteria were considered a nuisance, even parasitic. However, it is now recognised that each type of gut bacteria produces different chemicals and hormones, all of them essential for our health. Even some allergic or auto-immune disorders appear to be connected to an unhealthy or weak microbiota. It therefore pays to keep them healthy.

To that end, it is worth remembering that there is quite strong evidence that all they need is a diet fairly heavy in fruit, vegetables, and dietary fibre. Just like the famous Mediterranean diet.

Bacteria and fungi are everywhere. On our skin, on every utensil in the kitchen, and on every bit of food we eat. They also exist in the water we drink and every breath we take. The vast majority are completely innocuous, unless they penetrate our skin and manage to find a rich environment where they can proliferate, like our blood and cells. Many even protect us from various dangerous bacteria, viruses, and fungi. And a few of them are edible.

Edible bacteria and fungi exist naturally in many different foods such as yoghurts and sour milk as well as in various mixtures such as the Japanese miso. Probiotics (mixtures of various beneficial bacteria) are also now added regularly to natural foods as well, in an effort to improve our gut microbiota, which is often damaged by diets poor in vegetables and fruit.

The interesting thing from a physicochemical perspective is how these probiotic bacteria can manage to proceed almost unharmed through our digestive system and eventually reach our gut. It's quite an obstacle course, especially with the strong acid in our stomach. They apparently do this by shielding themselves using a thick mucus which reduces diffusion and even neutralises the acids long enough to allow them to reach our intestines alive. The nature of this mucus is not fully understood but it may be a type of fungus working synergistically with the microbes.

At the same time, edible fungi are also an important raw food in many dishes. Mushrooms and truffles are actually huge synergistic communities of fungal spores that have evolved to look like a single organism.[38] Structurally, they are held together by chitin, the same material as the exoskeleton of crabs, prawns, and lobsters. Yeast is the most common fungus we eat and a very important one, but it is only used for producing carbon dioxide for breads and is not directly edible.

In most cases mushrooms and truffles are eaten for pleasure (especially for their aroma), not for their nutrients. A lot of what they are made of, especially chitin, is largely indigestible, even by our gut microbes. After boiling, very little of the chitin dissociates, but frying does break down much of it, allowing for some limited decomposition and digestion.

Slow Food Versus Fast Food

As far as I'm concerned, food should always be cooked slowly and carefully. The ingredients must be added at the right moment and allowed sufficient time to blend and mutually diffuse to obtain the best result. Damaging chemical reactions due to high temperatures, which result in oxidation, dissociation, etc., must be controlled and kept to a minimum by cooking at the lowest temperature possible. Sauces need their own good time to polymerise and diffuse through, and reduction must be done very gradually. Nothing should be rushed. Meat must first be softened by gradually dissolving the collagen or breaking down the fibres before sensitive vegetables are added, and water must be allowed to diffuse through to make sure starches are softened, and so on. In general, slow food is a synonym for good food.

Normal cooking of a stew or a soup in a pot or a casserole in the oven is slow cooking with all its benefits. But is there a way to slow cook a thick steak of beef or salmon uniformly and succulently without drying it out? Of course there is. One method is called sous vide (French for "under vacuum") and it simply means that you place your steak or fish fillet in a plastic bag under vacuum[39] and cook it in hot water up to 90 °C for an extended period, i.e., a minimum of about one hour to get a medium-rare result for a thick fillet or steak. Sensitive fish fillets need lower temperatures, but one shouldn't

[38] Isn't this what every living being is? All living bodies are huge symbiotic collections of trillions of individual cells, after all.

[39] This is done with vacuum-sealing apparatus that uses plastic bags of high density polyethylene (HDPE) or polypropylene (PP), never polyvinyl chloride (PVC) or low density polyethylene (PE). We'll discuss the various types of plastics and their properties later.

cook anything at less than about 65 °C to ensure proper sterilisation of the food. Vegetables can be prepared like this as well and they lose none of their nutrients and aroma, as they tend to do in the pot, even if steamed. Boiled eggs can also be cooked this way to obtain a soft but not runny consistency. The main problem with this method is that meat comes out, well, boiled, so a steak still needs a quick searing in a frying pan or on a hot plate to brown it by the Millard reaction. As a student I used to boil sausages directly in their vacuum-sealed bag in the kettle with very good results.

On the other hand, "fast food" has come to mean burgers or fried chicken and similar, and refers to something quick and often a bit rushed. Not to mention dubious quality in some cases. Of course, not all such "fast food" deserves a bad reputation since many street foods all over the world are actually well prepared and carefully cooked. Fast food may be thought of as any type of high temperature cooking, from frying to grilling, and lots of physics and chemistry occurs in a frying pan as we've seen. But because of the serious protein changes occurring during frying and grilling, I personally prefer slow cooking in the pot or in casseroles.

Casseroles are an interesting case in point. They combine high temperature surface cooking with normal pot cooking. While the main body of the food cooks just like in a pot, the surface cooks a little like under a grill, exploiting the Maillard reaction to a gentle extent. In fact, there are dishes that cannot be made any other way. For example, anything that requires layering can only be made in the oven.

An example of casserole dishes which benefit from oven baking are various au gratin dishes where the cheese needs to melt and bind the protein (e.g., cauliflower with or without béchamel) before browning. Because of the low thermal conductivity of such foods (they contain a lot of air pockets), it is necessary to precook them (including heating the glass dish) before baking them. Exemptions are various vegan dishes like the Greek "briam" (a type of ratatouille) dish, where many different vegetables (e.g., zucchini with aubergines, carrots, potatoes, and tomato) are baked at high temperature to break up some of the tough fibres, and stirring is carried out regularly.

Lasagne with cheese topping or the popular Greek dishes pastichio or mousaka,[40] both with cheese-béchamel topping, are good examples of layered dishes requiring some precooking, followed by oven baking to ensure that heat penetrates sufficiently inwards. The dish must be assembled very quickly after preparing the ingredients (meat sauce, pasta or vegetables, and béchamel) and baked immediately so as to avoid losing too much heat.

[40] The name—but apparently not the dish—comes from the Arabic "mousaqqa" which means something like "pounded" or "cold".

For the same reason, namely the low thermal conductivity of most vegetable foods, many stuffed vegetable dishes are also cooked in the oven after some precooking. Two good examples are the dish "papoutsakia" (Greek for "little shoes"), consisting of stuffed aubergines topped with a cheese-béchamel sauce, and various popular stuffed zucchini or other vegetable dishes.

Be that as it may, in certain situations it does pay to speed things up a bit when cooking, or at least to accelerate cooking at specific times to obtain certain unusual results. We have already seen a few examples of this: crunchy fried chips, surface caramelisation, etc. The overarching objective in most cases is to form a protective skin which shields sensitive foods from desiccation or dissociation until they are properly cooked. Speeding things up in this case usually means frying, but it can also mean faster boiling, as when we want to evaporate an excess of water in a stew.

In most such cases, we create a physical barrier that is made of a different material from the main body of the food, for example, hard acrylamide on starchy potatoes or bread. Such physical barriers are dry and allow only a limited amount of heat into the body of the food. For example, by throwing potatoes in very hot oil we are aiming to rapidly form a hard crust which will protect the inside till it is properly cooked. If the oil is not hot enough then we tend to desiccate the potatoes and eventually even introduce too much acrylamide and oil into them—not a good idea.

In certain other cases, we want to encourage a certain additional aroma or taste, as in the caramelisation of onions (oxidative pyrolysis of the sugar) when we fry them lightly before introducing meat or fish or other vegetables into the pot. By the way, I like to caramelise carrots too, sometimes together with the onions. They tend to release a very nice aroma, and they also get sweeter by a transformation reaction.

The timing of accelerated cooking is important. For example, when we make a stew or a casserole, we want to go slow at the beginning until the meat (or other protein) is partly cooked but not dry, then increase the heat and fry it in oil in order to form a thin protective crust, before we add vegetables and slow the cooking down again to increase diffusion. At the end, we will probably want to increase heating again to enable thickening (polymerisation) of the sauce by adding some acidic substance.

Cooking with a wok on a hot flame is a case in point as it combines many of the above mechanisms. Here we are essentially frying very thinly sliced meat and vegetables in such a way that we encase the soft centres in a hard crust in a matter of seconds, while the large surface area of the finely cut ingredients and caramelisation of the proteins enhance the overall aroma. At the same time, we shake and stir the food all the time, which increases the contact between the ingredients and enhances diffusion. Because of the very short time (which should be) involved when we cook with a wok and the minimum amount of oil we (should) use, the overall effect is healthier than deep or shallow frying.

On Thin Ice—Chocolate Baked Alaska

Everybody loves ice cream (don't they?) and surely everyone loves chocolate. That's why chocolate chip ice cream is so popular. But what about a choc-coated melted ice cream bomb that explodes in your mouth?

It's one of the nicest summer treats you can prepare (and an impressive dessert) and it all relies on microwave energy being able to penetrate dry foods as easily as if they were not there. You'll need a few small ice cream balls (or bells) with dark chocolate coating and a very cold plate. If you cannot find them in the shop, you can easily make them by dipping scooped balls of ice cream into molten dark chocolate and re-freezing them. Now, put a few of them on the cold plate without touching them and immediately microwave them for about 15 seconds. The result is a cold, hard chocolate ball containing molten ice cream that explodes with one bite in your mouth. The MW energy penetrates the cold hard chocolate—which contains very little water—as though it wasn't there, and heats up and melts only the ice cream inside, with a delightful result.

The same thing happens when you heat up any food with a crust but a soft interior such as a pie or a dumpling. In fact, microwaves heat up the interior rather than the exterior of any food, even a soup. The reason is that, just like any form of light, they can also be reflected, and a shiny, wet exterior reflects light better than a dry crust. When reheating a soup or a stew in a MW oven, it's a good idea to stop every thirty seconds or so and stir it to allow heat distribution. Alternatively, reheat it at half power for double the time[41] and let thermal conductivity do the job of distributing the heat evenly in the food.

Finally, when I make the original baked Alaska, I aim for a similar effect, but instead of chocolate, I cover very cold ice cream balls with meringue and put it directly under a very hot grill for a few seconds to set the meringue (solidify the albumin by cross-linking the macromolecules) and

[41] "Half power" in MW ovens does not mean exactly that. It just means full power applied intermittently (i.e., "half the total energy"), as we'll discuss later too.

slightly brown it. This gives a crusty protection with a soft ice cream centre. Nice, but I prefer the chocolate version.

Stretchy Dough

There are lots of occasions when we want to create unusual cakes or breads or pies, and for this we need to produce a strong, stretchy dough. What is the physics behind this?

The basic ingredients are deceptively simple: flour, water, and a lipid—butter or oil. But if you combine them all together at the same time, you often get a crumbly, or at best non-elastic dough. Indeed, if you add water first, you get crumble. The reason is that the water combines preferentially with the flour, closing all pores and excluding the oil, making the dough crumbly and poorly homogenised. Because water is a polar liquid while oil is a non-polar, hydrophobic liquid, the two will always try to exclude one another. Once the flour has absorbed the water, it will not be able to absorb the oil, which is critical for the elasticity.

What is needed is a binding agent and this is the job of the flour, up to a point. The only way to ensure a strong, elastic dough, is to first mix the flour very well with the oil (or molten butter), until it is well bound (salt helps a little) and then introduce the water gradually. The oil will become even better distributed and will not easily separate from the flour as it tries to avoid the water. At the same time the water, being polar, will adhere very strongly (both electrically and mechanically) to the flour, and this will ensure that the whole dough gets very evenly homogenised. It's a synergistic effect. For best results you need to leave the dough in the fridge well covered for at least one hour to enable sufficient diffusion and to stop the lipid from melting and diffusing around in the dough. The end result is a very stretchy, very elastic dough you can use for pitas, pizza or pies. If you use whole wheat flour as I do (about 50:50, and up to 70:30, which gives tasty results but is trickier), you need to use a little more lipid to start with as the fibres tend to lock-in quite a bit of the oil.

When making croissants or other kinds of very flaky pies or fibrous bread (such as the "tsoureki" sweet bread popular in Greece), then even the above basic stretchy dough is not sufficiently elastic. In such cases, we need to add a slightly beaten egg, whose albumin aids in bridging the oil and the water even better because of its electrical affinity for both. The egg must be added to the flour first and mixed well, and then the oil or molten butter is added and mixed in well before the water and the salt and any other aromatics are added. After an hour in the fridge, this dough is extremely elastic and can be folded over many times (oiling the surface every time) *without kneading* to create a croissant or puff pastry dough. Even though it looks like a single mass, when baked these folds will separate out into multiple layers because of the physical action of the intervening lipid, just like many layers of fyllo put together. In fact, even the baked product behaves quite elastically because of the relative sliding of the multiple layers. It's the same principle that allows a laminated brittle material to bend without breaking.

Wholesome Food

The benefits of using whole grain flour (or rice) cannot be overstated. Apart from the beneficial effects on our own digestive functions, the husk and fibres are the ideal slow food for our gut bacteria, keeping them healthy and producing critical hormones and other compounds that our own cells cannot. In addition, the use of whole grain flour delays the conversion of starch to sugars in the body, allowing for more uniform and longer acting metabolism while reducing any sugar spikes after eating. It appears also that gluten from whole grain flour has a much reduced allergic effect on people with gluten sensitivity.

Concerning the flour for breads and cakes mentioned above, I actually use a mixture of whole grain and white (refined) flour at a ratio of more than 50% whole grain flour, usually with added cracked or floured oats, barley, and various seeds. Ideally, I'd like to use 100% whole grain flour, but to get a nicely raised bread I'd need a lot of yeast (at least triple the amount)which alters the taste a bit. As we saw, this is because whole grain dough tends to leave open pores and much of the carbon dioxide escapes early before it has a chance to raise the bread. Indeed, this situation gets worse on exposure to a high temperature in the oven. For this reason, the white flour (about 30–40% in most cases) serves to close the pores and allow to raise the dough without using excessive yeast. Commercial whole grain breads usually contain a much larger proportion of white flour, up to 80%, in order to raise the dough further and produce a light, fluffy bread. Of course, I'm referring to real bread, not the sticky industrial glop that passes for bread and is not made with yeast (the sliced "bread" for toast).

Whole grains (of wheat, barley, or any other cereal) include the husk of the seed which, being almost completely fibrous, has very different water absorption properties to the inner seed. In fact, the husk has evolved in order to protect the seed from premature dissociation or decomposition, as well as attack from moulds (omnipresent in the air), so it must first be broken down by extensive cooking before the seed itself can start cooking, something almost impossible at reasonable temperatures and in reasonable times. Even during soaking, the husk absorbs water by diffusion and swells without allowing too much of the water to diffuse inwards. Any mould spores tend to be stopped and broken down at the husk as well. For this reason, to use whole grain flour for baking we have to grind the whole grain, which breaks it up and exposes the seed. Even then, whole grain bread requires much more yeast, longer baking times, and slightly higher temperature than ordinary flour.

A good example of how different (and tasty) whole grain bread can be is the well-known vollkornbrot German bread mentioned earlier, made using one of the original, ancient wheat species, the einkorn wheat, a little sugar, and beer. One can also make it with a little molasses instead of sugar, which (with the beer) is used to increase the metabolism and carbon dioxide output of the yeast.

Whole grain rice is another healthy food for the same reasons as the whole flour, and in this case, there is no gluten. Even the sticky "glutinous" rice used in Chinese cooking actually contains no gluten at all. Because of its husk, whole rice (sometimes called brown rice) requires at least twice the amount of time to cook at boiling temperature to hydrolyse the husk before the seed can start cooking.

Explosive Food

Earlier, I mentioned that slow food is generally the best food, but there are times when we need to speed things up a bit. That's when we sometimes get explosions in the kitchen, some of them spectacular.

In fact, it's not that unusual for some food to explode during cooking, mainly as a result of steam pressure building up in an enclosed space. Sometimes we even try to create some controlled explosions, for example, to create foamy foods or other interesting concoctions. Let's look at some examples of both uncontrolled and controlled explosions.

Everyone knows that you shouldn't attempt to boil a raw egg (in the shell) in the microwave. The shell is mainly made of calcium and magnesium carbonate, which are non-polar and dry. This means that they are completely transparent to microwaves and don't absorb any energy. On the other hand, the water inside the egg (more than 80% of the total mass) readily absorbs microwaves and immediately starts heating and evaporating.[42] Very quickly (within a few seconds, well before the egg white has time to heat up and coagulate and solidify), the steam pressure will build up beyond the internal stress that the shell is designed to withstand and it will explode quite violently, making an absolute mess of the microwave oven.

In fact, even an open egg without the shell can explode in a microwave oven, as I have found out not infrequently when cooking very fresh eggs. A very fresh egg yolk is covered by quite a strong lipid membrane,[43] made of non-polar molecules and therefore transparent to microwaves. That's why a fresh egg yolk sits almost spherical in a bowl. In an older egg, however, the lipid structure will have been weakened (hydrolysed) by the water inside and outside, and the yolk sits flatter in a bowl because the membrane is weak and easy to bend and break. Microwaving a very fresh egg (the fastest and healthiest way to cook it—no need for any oil at all) needs to be done at a low power setting to avoid overheating and exploding the yolk, whereas an older egg can be cooked quite safely at the maximum setting.

Leidenfrost's dancing drops

Have you ever noticed that a single small drop of water on a hot plate will dance around and take some time to evaporate completely? But if you spread it out it will disappear immediately?

This happens because some of the water on the surface of the drop evaporates immediately and, once it does, the ensuing steam insulates and cushions the drop from the hot plate, making it jump and dance around. This is called the Leidenfrost effect, after its discoverer.

The same protective surface-localised evaporation effect allows you to dip your hand in liquid nitrogen (at a temperature of − 196 °C) or even a cold hand in molten lead (+ 327 °C) for half a second or so without any harm. But please definitely don't try it at home!

Similar explosions can occur with all sealed vegetables in the microwave oven, especially potatoes (some fresh ones can crack or even explode within 2–3 min) and aubergines (egg plants).

A dangerous explosion can happen whenever high energy input is not balanced by an equivalent energy release. Microwave ovens are not the only potential culprits. A dangerous situation can also occur if frying oil is heated up too much and we add a load of wet potato chips. The surface water on

[42] All materials evaporate, even at much lower than boiling temperatures, because of the random energy distribution of the molecules.

[43] It's the membrane that sperm need to penetrate and fertilise the egg. As soon as one manages it — not a mean feat—the membrane reforms by cross-linking to form a hard, impervious structure.

each potato will heat up and boil almost immediately, creating thousands of steam bubbles which join up and expand rapidly, suddenly pushing the whole oil mass upwards. If the oil overflows the pot, it will catch fire, and it can even explode if it seeps between the pot base and the electric hob plate. If we are not sure whether the oil temperature is correct, it's always a good idea to add chips in gradually. Then, if it's too hot, it has a chance to cool down a bit before adding the rest.

An explosion can also occur even inside a pot full of otherwise gently boiling stew with rice. This can happen if there is no stirring, or inadequate stirring, and the rice (or flour, as in bechamel sauce) settles on the bottom of the pot, creating a seal beneath which some liquids continue boiling. Because rice is a good thermal insulator and seals well, little heat is conducted through it, and all the while, steam will build up underneath. Eventually, that layer of rice may suddenly rise or even explode inside the pot. So it's a good idea always to stir the rice frequently if you do not use a rice steamer.

Other explosions can occur because of the flammability of various food stuffs, such as icing sugar, flour, etc., in a very dry environment. All such foods have such a small particle size (nanometre-size) and, as we discussed before, their very large specific surface area means that they can readily react with oxygen to cause a fire. Even simply throwing dry flour or icing sugar directly on a gas flame or a hot stove or frying pan can start a fire.

On the other hand, we sometimes cause controlled explosions in the kitchen for particular baking creations. We may even consider a cake to be an assembly of thousands of small explosions in which baking powder releases carbon dioxide in closed cells, thereby causing the whole cake to rise.

In another example, during the final stages of making a thick sauce like béchamel, bubbles burst through the top like small explosions, telling you it's time to stop heating and remove the sauce from the hot plate.

Finally, frying often causes minor explosions on specific foods. Calamari and similar sea foods, even after cleaning, are covered by thin but strong membranes containing water which persists even after coating them with flour. When frying in oil, the trapped water can overheat and it's not unusual for small explosions to make a mess of your clothes. This means an apron is highly recommended in such cases. Vegetables with thick external skins, such as aubergines, can also give the same explosive result during frying.

Overcooking Trouble

I've talked a lot about the chemical conversions that occur during frying at high temperatures or when exposed to fire. The added energy from the fire transforms many compounds in the food by pyrolysis (Greek for "dissolution by fire"). Starch converts to acrylamide, meat after the Milliard reaction converts to various toxic chemicals,[44] and fruit and other sugars caramelise[45] before charring. Normal heating and cooking do not necessarily produce all these chemicals. It's only when water has been evaporated away and the proteins are exposed directly to the hot plate or fire. In all these cases, the atoms absorb too much energy and vibrate violently enough to cause the proteins to quickly uncoil, be damaged, and then reform. Even when removed from the energy source, the damaged proteins (or separated amino acids) can reform into different chemicals or, more likely, remain denatured. For this reason, the golden rule in cookery is to stop when a little water still remains (except when making soup). This point may not always be obvious, of course, but we can get a good idea by watching the behaviour of the food. In the pot, all free water[46] has evaporated and the stew is ready when boiling becomes explosive or the food starts sticking to the bottom, as can be felt by a wooden spoon. In the oven, a pie or a casserole is ready when the former becomes golden and starts shrinking away from the sides of the tray and when the latter's surface boiling becomes explosive. In all these cases these observations indicate that the free water has evaporated and the temperature is already climbing above 100 °C.

If the protein molecules persist on the hot surface or fire, then all the side bonds of these proteins and amino acids will be destroyed by pyrolysis. The hydrogen and oxygen ions will happily combine to create water and the various nitrogen and metals ions will also be oxidised into gases or converted to foul smelling compounds such as sulphates, nitrates, etc. All that will be left is the skeleton of the proteins, consisting solely of black carbon atoms.

Overcooking food doesn't just alter its chemistry and produce toxic chemicals. It alters its taste and makes it indigestible as well as dangerous. One might as well eat a few pencils—that's also carbon.

But what can one do when asked to prepare something "well done"? My own immediate reaction is always to refuse—I don't want to destroy the food

[44] Heterocyclic amines (HCAs) and polycyclic aromatic hydrocarbons (PAH) among others. All of them are dangerous, with strong indications of carcinogenicity.

[45] Releasing the "nutty" smelling chemical diacetyl.

[46] Free water is water that is not bound up with other chemicals or mechanically bound within cells and other structures in the food.

and poison my guests. The most I will ever cook food is to a brown colour, allowing some of the transformation reactions to take place and only on the surface.

I can't honestly believe that anyone really enjoys overcooked food. One possible reason some people ask for "well done"[47] is because they are worried about the possible presence of bacteria, especially in meat. In that case, if you cannot trust the quality of your meat, boil it, or better, microwave it first for a minute or two until the internal temperature reaches 65^0C, which essentially sterilises the food, before frying or grilling it for a short time to give it colour. For the same reason, if you cannot be sure of the freshness and quality of your eggs, cook them to at least 65 °C for a minute or two, at which temperature they only just start denaturing and solidifying. By the way, a soft (almost runny) omelette like the tortilla Espanola (with potatoes) can be made easily and safely (properly sterilised) by beating the egg slightly in some milk (but not completely breaking up the albumin proteins), pouring it in a hot, oiled frying pan with the already fried potatoes, and stirring continuously only during the first 15 or so seconds to make sure all the egg has been well heated up before becoming fully set. Finally, put it under a hot grill for another half a minute for the perfect soft tortilla omelette. I often add a bit of cheese, origanum and perhaps some thinly sliced meat.

Shaken, Stirred, Beaten, Kneaded, or Whisked?

Mixing and blending of ingredients is one thing, but sometimes we need to introduce more mechanical energy to enable better mixing or to force ingredients to bond. However, the amount of energy we put in has to be controlled, otherwise the result will end up all over the place.

The lowest energy mechanisms of mixing and blending are simple shaking or stirring. We use them when the ingredients must simply come into contact with each other and we must avoid too great a chemical reaction between them. The weakest is shaking, which leaves solid food intact in the pot or casserole and only circulates a liquid around it. It is used when we introduce an emulsification agent that must circulate, but is very sensitive and cannot be stirred too much because it can easily curdle. A lemon–egg emulsification agent poured on stuffed vegetables laid out in rows in the pot is a good example.[48]

---

[47] What a misnomer. It should rather be called "badly done" or overdone or destroyed food.

[48] Greek vegan stuffed vine leaves (dolmadakia) or filled cabbage leaves (lahanodolmades) are good examples.

Next up in energy input, stirring, introduces more energy into the food. It forces the ingredients to come into contact with each other and with the hotter base and sides of the pot. Stews, ratatouilles, and simple soups are good examples where the cooking temperature should be controlled. Just letting the soup boil on its own and hoping that it will result in a good blend does not work well because of the high temperatures on the sides and bottom of the pot and the internal circulation currents set up in the pot during boiling.

Beating a mixture of liquids introduces even more energy and is mainly used in cases where the liquids are largely immiscible and require energy to get them to blend well. Beating whole eggs helps to break down the lipid membranes and mix them with the protein and the albumin. But beating can go too far and give the wrong result. If we are making an omelette, we need the lipids and the albumin to remain mostly intact so as to form the mechanical structure keeping the omelette fluffy during frying. Mixing in a little full-fat milk while keeping the beating to a minimum, preferably using only a fork, is best. If on the other hand we want to prepare a smooth sauce like béchamel, then we want all membranes and albumin to be fully mixed with the protein, so the beating (or better, whisking) must be vigorous, whether with a hand or electric beater. Intense initial beating of sugar, eggs, and butter is also necessary when we prepare a cake mixture, otherwise the albumin will fail to emulsify the oil with the water in the cake. But the flour and other agents must be added by soft mixing or folding.

Bread dough presents special problems as there is not enough water to give good bonding with the gluten without long and strong kneading to mix in the water and blend the ingredients. If we want to add egg to the dough then it needs to be beaten well first and then added to the initial mixture before kneading well to bond with the gluten–water mixture.

Finally, making frothy or fluffy creams (Chantilly) requires whisking or beating so that the albumin bonds with the water in the protein and forms intricate semi-rigid networks containing air. This is only possible with very vigorous beating and whisking which provides a lot of energy. Extra icing sugar (if desired) should only be added after beating, otherwise it will immediately react with the albumin, breaking the polymer chains and not allowing any thickening.

Making Food Last

We've already discussed the sterilising and preserving effects of salt and sugar. Both work mainly by desiccating the protein, thereby removing the water needed for parasites to grow. But they are not the only means of ensuring that moulds (fungi) and bacteria will not multiply or thrive in our prepared foods. And if you, like me, love left-overs, we need to make sure the prepared food is as uncontaminated as possible before eating it. Most left-over soups, stews, and casseroles taste better the next day or two (as long as they have been stored in the fridge) because diffusion of nutrients continues, even in the fridge. This ensures a better homogenisation of sauces as well.

The main parasites that invade all cooked foods—and all foods in general—are thousands and thousands of different types of moulds. Mould spores are everywhere—in the air we breathe, the water we drink, and the food we eat. There is absolutely no way to avoid them, no matter how clean we keep our kitchen. In fact, nearly all cases of "food going off" involve moulds of some sort. Many fruit and vegetables are susceptible to moulds to varying degrees, especially if their protective outer skin has been damaged somehow. The way they work is by injecting special enzymes into the fruit or vegetable or meat which break down everything—proteins, sugars, fats—into their smallest units (amino acids or even simple molecules) and then reabsorbing them for their own metabolism. In order for the enzymes to be injected and work their damage, they use water as the diffusion carrier.

Interestingly, the way moulds grow and spread within the food is often different in various species of fruit and vegetables. For example, a small scratch on most apples will quickly (in a matter of 30 min!) spread through the flesh and the characteristic mouldy smell and taste will contaminate the whole apple very quickly. This is because apple cells tend to communicate with one another, and their cell membranes are easy to break down. The enzymes sent by the mould break down the cell structure and denature all proteins to a liquid mixture of amino acids and water that can be absorbed and metabolised by the mould. On the other hand, the cell structure of pears and peaches is quite different, with limited diffusion paths and hardier walls between cells, so a mouldy area on a pear or a peach can be cut out and the remainder eaten safely, if you catch it early enough. The same is true for oranges and mandarins and other similar fruit. If any section is infected, in most cases the rest of the fruit is unaffected, unless it's the green, hairy type of mould attacking the outside of oranges or lemons. Because those particular types of mould extend long tendrils inwards, the whole fruit is quickly contaminated. Vegetables also vary widely in their response and sustainability with regard to mould infection and we'll discuss their response below.

We breathe in and swallow mould spores all the time, but it's rare to get a fungal disease.[49] We should be thankful that our immune system has evolved over millions of years to be very well focused and expert in killing them, so we hardly ever notice them.

Most types of food are susceptible to bacterial or fungal contamination, except dry foods where water has mostly been removed by high temperature frying or grilling. Dried herbs, vegetables, fruit, pasta, rice, and nuts (if mould-free to start with) are also mostly impervious to any bacterial contamination. Drying meat or fish (protecting and drying with salt and spices) is a common method of preserving food and it has been known for thousands of years.[50] Smoking ("curing") has also been used to preserve different protein and fat-rich foods, because wood smoke and raised temperatures kill most bacteria and gradually dry the outside of the food. Alcohol is another agent that destroys bacteria by dissolving their membranes. This is what preserves wine, beer, etc. Finally, most modern processed foods contain various chemical preservatives that work by destroying bacterial membranes, although they do change the taste and may not be as safe as we wish.

Cooking well essentially sterilises all food. No mould or bacteria[51] can survive boiling temperatures in a pot or the scalding temperature in a frying pan or an oven.[52] However, in general, boiled vegetables and fruit are much more susceptible to mould contamination later since half the job of breaking down the cell structure has already been done by the cooking and moulds can more easily gain a foothold and spread. Any contamination of jams, sauces, etc., by mould spores (possibly) or bacteria (very rarely) will thus happen after the food is cooked and has cooled down and only in the presence of free water.[53] If there is any free water on the surface of the food, mould spores will immediately grab the opportunity to embed themselves and start growing and multiplying, aided by the nutrients below. For this, they need

[49] But not impossible. Immune deficiency may allow fungal infections, but most are relatively easy to treat.

[50] Interesting examples are "biltong" from Southern Africa, "kilishi" from Nigeria, and "beef jerky" from the USA.

[51] There are some extremophile bacteria that can survive boiling, but they are extremely rare and certainly unheard of in a clean kitchen.

[52] All bacteria (microbes, viruses, spores, parasites) are enclosed in membranes made of various lipids which dissociate easily above 65 °C or in alcohol or under exposure to sunlight, resulting in their death.

[53] It is mainly moulds we should be concerned with in the kitchen, since bacterial contamination of cooked food is very rare unless it pre-existed in the ingredients and cooking was incomplete.

water and some air, so we need to exclude either or both of these to stop moulds growing.

Breathing water

Water in the form of extremely small droplets and steam is easily dispersed and held in air in a humid environment. We specify the relative humidity (RH) as the amount air actually holds divided by the maximum it can hold at each temperature. At 100% humidity, the air cannot hold any more and condenses everywhere, so nothing dries.

It's remarkable how much water air can hold. The maximum amount depends on the temperature, but in a typical kitchen at 30 °C it can hold up to 30 g of water per cubic metre of air. If the relative humidity is 80% the air holds about 24 g of water, so we breathe in about a teaspoon of water every minute.

Because we rely mainly on sweating to cool down (by losing latent heat via the evaporation of sweat), a simple dry thermometer reading is a very poor indicator of comfort in humid environments. Much better is the "wet-bulb" thermometer reading, which takes into account the evaporation. The "dew point" (the temperature at which water condenses) calculation is also a better, but a bit more extreme indicator.

This means that a way must be found to stop any contamination, and the simplest way is to reduce all free water on and in the food. Freshly cooked food must be cooled down completely before covering and storing in a fridge, to reduce the energy available for parasites to grow. Cooling down for an hour before storing reduces any condensation that may form on the underside of the container lid because of the difference between the temperature inside the container and the temperature inside the fridge. As an added precaution, after a few hours in the fridge, carefully dry out the condensation from inside the lid. Condensation is also reduced if you minimise the amount of air in the container before closing it.[54] Freezing of cooked food should be done only after a day in the fridge and removal of all condensation. "Freeze drying" of flaked or powders food in factories is done at very low humidity and at very low temperatures which ensures almost zero free water.

Condensation of steam is actually present everywhere in the kitchen, during and after cooking, whenever steam meets a cooler surface: on the underside of the pot cover, the walls, the windows, etc. If the extractor fan is not extracting the steam properly (see later), we'll also have condensation on the underside of the fan, which can easily drip back into the pot. This is

[54] Some condensation will occur in a humid atmosphere anyway, even if you cool down the food to room temperature. Transparent covers help to monitor any condensation so that you can remove it in time.

because the fan casing is cooler than the surrounding air so any steam will easily condense on it if there is insufficient extraction. In general, condensation is almost distilled water so it is almost sterile and cannot do any harm if it falls back into the pot from its lid during cooking.[55] But not condensation from the extractor above.

The pressure of air

The pressure of the kilometers of air above us (and everywhere around us) is very high. At sea level, it is about 100 000 N/m^2 (100 kilopascal). That's equivalent to 200 kg squeezing your open palm! We don't feel it of course, because we are adapted to it and because it squeezes us equally from all directions ("isostatically").

Many sensitive foods are vacuum packed, which helps to keep them fresh and makes them last longer. They are usually packed at only about 1% of atmospheric pressure, so the atmosphere squeezes and seals the food with a pressure of about 99 kPa. That's why it is so difficult to open a vacuum-packed jar or packet.

How else can we reduce the amount of free water in contact with spore-filled air in food? For soups or stews or boiled vegetables, an excellent way of doing this is to add a thin layer of oil, something which happens automatically if we use a sufficient amount of oil in the food when cooking. The portion which is not bound in the sauce will sit on top of the food since it is lighter. All oils have a lower density than water so they float on top, forming a barrier that is toxic and impervious to most bacteria and moulds (oils break down their lipid membranes). Almost nothing grows in oil so this is a quite safe protective barrier, although it does oxidise over long periods of time. In many Mediterranean homes and restaurants today (and in ancient times as well), nearly all pot and oven dishes are made with a slight excess of olive oil, both for taste and for preservation. Olive oil is also an excellent preservative for pickled foods, olives, and all types of dips and sauces. For added safety, pickles and vegetables should be kept at low pH (less than 4) with some vinegar, because moulds and bacteria cannot grow under such conditions.

Moulds, like all life, depend on water to grow, but many depend on the availability of oxygen as well. Reducing or removing the air from a container not only reduces the available water but stops metabolism of parasites. The food industry utilises this principle widely, and vacuum-packed foods are

[55] Apparently, some people believe that drinking distilled water is healthy. It is nothing of the sort and actually quite dangerous in sufficient quantities as it destabilises the sodium/potassium balance across cell membranes.

now the norm if extended shelf life is needed, for example, for pre-packed cheeses, processed meats, etc., which use special multilayer plastic films. Be that as it may, all plastic materials are ultimately pervious to oxygen, so even vacuum-packed foods cannot be preserved indefinitely. The oxygen transmission rate (OTR) is an important quality parameter for plastic films used for packaging. They need to have an OTR thousands of times lower than ordinary cling-film. Simple kitchen cling-film (now made mainly of low density polyethylene, LDPE[56]) cannot keep a vacuum, as it is quite pervious even to water and must be used only as temporary packaging. More expensive aluminium-coated plastic films have almost zero OTR and are used for aromatic or sensitive materials such as ground coffee, chocolate, croissants, etc.

The vacuum packing method cannot be used for ready meals, nuts, dried fruit, etc., because it would dry them out completely.[57] Instead, food manufacturers pack ready meals in a nitrogen atmosphere at atmospheric pressure, thereby excluding oxygen, but the OTR of the barrier must still be very low to ensure oxygen doesn't leak in by exchanging with nitrogen molecules. They also flash freeze prepared dishes very soon after producing them to reduce the possibility of bacterial or fungal growth.

Modern food cans (or tins) need to preserve food for years and are made of aluminium or tin-coated or chromium-coated steel, but metal contact with food is generally avoided by coating the inner surface with a special plastic barrier with high OTR.[58] The barrier needs to have other properties too, such as mechanical tear strength, wear resistance, temperature and chemical stability, etc. Glass packaging, fully sterilised, is used for all acidic or liquid foods, such as sauces, jams, etc., which are usually vacuum packed too.

When making marmalades, jams, or sauces at home, it is important to sterilise the glass jars as well as possible, otherwise moulds will find a way to grow on even the tiniest amount of free water, e.g., condensation. The classic method to ensure that is to boil them standing upside down for an hour, but you still can't be sure you won't pick up some spores afterwards. There is a simpler method. Just add a few drops of alcohol (vodka or ouzo will do fine) into the jar, close it tightly, shake it and leave it for an hour while you are making the jam. When ready, pour in the hot jam, fill to the brim and close the jar tightly. As it cools down the little bit of air trapped

[56] They used to be made of softened poly vinyl chloride (PVC, also used extensively for plastic containers until recently) until it was discovered that the phthalates used to soften PVC are very dangerous.

[57] Water would readily diffuse out of the food onto the surface until it reaches vapour balance.

[58] It used to be bisphenol-A-based plastic (BPA, an epoxy), but now other plastics are used, as BPA has also been found to be very dangerous.

inside will contract creating a moderate vacuum[59] containing a little alcohol vapour. Nothing survives that and the jars can be stored in the cupboard for months.

But how can you tell if something has been attacked by a mould? As I discussed previously, there are literally hundreds of thousands of different mould species and they all just need water and a few nutrients to grow. Many appear hairy (like the green mould on lemons and oranges, and also the invasive moulds on yoghurts) others are just slimy. Many, but not all, have a characteristic "mouldy" smell. But they all have one thing in common: they inject special enzymes which break down fats and proteins—even the hard, skin proteins of fruit—into their very basic constituents in order to be able to absorb and metabolise them, and they also liberate water in the process. So if you see a very soft, watery patch on a vegetable or fruit which is broken and has lost its normal smell—even if there is no obviously visible mould or slime around—it is usually the result of a mould (or microbe) and you can try cutting around it before eating the rest, provided the rest smells and looks normal.[60] Tomatoes and cucumbers are susceptible to moulds but they are rarely affected beyond the immediate vicinity of the attack. Over-ripeness sometimes looks like that, but the fruit or vegetable still has a completely normal smell. In this case, it helps to refrigerate it to help keep longer.

Fermented Shark, Anyone? Heat-Free Cooking

Have you ever heard of kiviak (or kiviaq)? It is an Inuit delicacy from Greenland. It is made by stuffing many whole Auk birds into a fresh seal skin (with the fat layer still intact), sealing it, and leaving it in a cold environment for months. In winter when food was scarce and difficult to obtain, the Inuit would actually survive eating kiviak and similar fermented foods. Very successfully preserved even if not easily enjoyed.

Fermentation is the only exothermic process which does not require oxygen. Nearly all heat-producing processes, such as metabolism, on Earth need oxygen, but fermentation does not (neither do nuclear and atomic processes, but that's another story). A gas fire burns gas in oxygen exothermically and the same is true with a wood fire. Without oxygen there is no fire.

[59] A half-filled plastic water bottle in the fridge will crumple and collapse for the same reason. This method is of course used by many industries to vacuum-pack jams and similar foods. When you twist the jar open (difficult, as you are working against atmospheric pressure—use a blunt knife under the rim), you hear a click telling you that you have "broken the vacuum" and let some air in.

[60] The general advice is that very young children, along with older, vulnerable people and everyone with weak immune systems, should be extra careful with any moulds.

That's how fire-fighting liquids or foams work, by excluding oxygen from the fire. In fact, the more oxygen you add to a fire, the higher the temperature and the stronger (and faster) the fire. There is some evidence that millions of years ago the atmosphere contained more oxygen than now (up to 28% compared to 21% now). This would have allowed fast and strong growth of plants (and animals) in wet climates. When the climate changed though, huge fires swept through everything, but many plants had little access to oxygen and produced half-burnt wood (similar to brown charcoal). Over many millions of years, this converted to vast deposits of coal and gas and petrol that became the fossils we have been using over the last 150 years to produce energy (and pollute the atmosphere). Electricity is unfortunately still largely produced by coal, petrol, or gas combustion with oxygen. Hopefully, the next few decades will see a complete decarburisation of energy production.

But let's get back to fermentation. Since it is an exothermic reaction, the heat it produces can actually cook the fermenting food very slowly, even in the absence of oxygen. I haven't had the dubious pleasure of eating kiviak yet (and I'm not sure I would want to, hearing about it from people that have tried it), but it is a fascinating process. Fermentation also happens in the body, but it produces much less energy than oxygen metabolism by our mitochondria.

Left-Overs Taste Better

If your cooking and storage methods are correct, you'll have no problem keeping a meal in the fridge for a few days and enjoying it again. It makes both financial and culinary sense. As I mentioned earlier, reheated fresh food (in the microwave oven of course) often tastes better because of further co-diffusion and co-absorption of ingredients and nutrients. Even at the low temperature of the fridge (about 4 °C, if it is working as it should), ingredients can diffuse around and some reactions can even proceed further.

The danger with reheating food from the fridge is the possible growth of moulds and, in very rare cases, bacteria, due to the presence of free water at the interface between the water and the air. Moulds almost never grow inside the food (unless there are air pockets with some water), but always on the surface, where they will be clearly visible.

In any case, it's probably a good idea to eat left-overs within a few days at the most. As a rule of thumb, foods containing a creamy sauce (e.g., béchamel) in a full, tightly closed container can keep happily for up to 3–4 days in the fridge, while freshly stored meat or vegetable stew or casserole or tomato-based sauce can easily keep for 5–6 days as long as free water is kept to a minimum and the container is as full as possible and well closed. Always choose the smallest container possible that will fit all your left-over food, as this minimises the amount of air above it. Simple boiled foods like pasta or rice can keep even longer if fairly dry. Fried or grilled foods, being even more sterilised by the high temperature frying, can keep for up to a week, but they will probably dry even more by that time.

When reheating left-overs, it's always a good idea to let them reach room temperature before reheating them thoroughly in a microwave oven, stirring occasionally as necessary to homogenise the food and distribute the heat evenly. The thermal conductivity of food is very low and heat will take a long time to distribute itself throughout by conduction alone. Using a glass container with a well-fitting lid will give you very good results. It's not a good idea to use plastic containers (except PP) in the microwave oven for reheating, as it is not fully established that their constituents do not leach out at the hot spots.[61] A food-grade PP container is as good as glass as long as you don't overheat the food.[62] Keep the lid only slightly open to allow steam to escape, but not so much that the food dries up.

Reheating in an oven is never a good idea as it will just heat up the surface of the food, while the inside will remain cold. If you are forced to use an oven, cover the food well with aluminium foil and place it in the centre of the oven. It will eventually heat up your food, but you'll wait a long time and waste a lot of electricity or gas. In the absence of a microwave oven, pasta and rice can be reheated fairly effectively by adding a little water or milk and heating them in a pot, while stirring continuously. The liquid will pick up heat from the base and help to distribute it throughout the food more or less evenly.

Always Store in a Cool Dry Place?

The general advice "store in a cool dry place" is good practice, of course, but it's not enough for food. As soon as food has been properly cooked, it is completely sterile, but it does not remain so for long at room temperature, and needs to be well protected to avoid contamination by bacteria, and even more so moulds. Apart from everything noted above about removing

[61] Even with a rotating table, some microwave ovens form hot spots on the food. The ones with convex or concave reflectors are better as they distribute heat more evenly.

[62] Polypropylene (PP) can be used up to about 100–120 °C, but microwave hot spots can reach higher temperatures. Glass presents no problem at all, even at much higher temperatures.

free water and air, we enhance its longevity by reducing the energy content of the food, i.e., by substantially reducing its temperature. The lower the temperature, the lower the rate of all chemical reactions, including natural dissociation reactions and bacterial and fungal growth. The underlying physical mechanism is always the energy availability of the molecules. Below the freezing point of water, the diffusion of molecules through ice is heavily curtailed, reducing reaction rates even further. Below about − 18 °C, chemical reactions have all but ceased and protein-rich foods will remain as they are for weeks or months. All good freezers (with at least 3* indication) work at − 18 °C or a bit lower. Deep freezers work at even lower temperatures, but offer only limited additional benefit.

All protein-rich foods—raw as well as cooked—should be stored in the refrigerator whenever the ambient temperature is above about 20 °C, even if they are going to be used within the next few hours. This is particularly important for fish and fatty foods, which are more susceptible to bacterial contamination and growth.[63] In fact, even if you'll use the meat or poultry or fish after just a few days, it's better to place it in the freezer and defrost it in the fridge for 12 h before cooking. If the kitchen temperature is less than 20 °C, then you can defrost it outside, carefully covered.

Raw game meat needs particular attention since most wild game meat (and sometimes river fish) is riddled with parasites. If very fresh, game meat needs draining of all blood, and then it must first be left hanging in the open air (sunlight is optional) for at least 24 h, protected behind a fine mesh.[64] Afterwards it should be placed in the freezer at about -18 °C. At least 2 days before use it should be thawed and marinated while covered in the fridge with plenty of vinegar, onion, and oil (as well as wine and other herbs, according to taste) for at least 2 days turning, over occasionally. Deep diffusion of the marinating liquids will break down their cell walls and kill all parasites and their eggs deep inside the meat, and also initiate the breakdown of strong fibres within the meat, which helps to tenderise it.

While discussing preservation, I'd like to add some comments about "expiry dates" on foods. I have struggled to understand the real purpose of expiry dates or "best before" dates on so many food items. Generally, well-stored fresh foods do not lose their nutrients or become infected for many days. More often than not, perfectly eatable[65] food gets thrown away because

63 Lipids dissociate much more quickly than proteins, even under bacteria-free conditions, as their molecular hydrogen bonds are weaker.

64 You can ignore this for butcher-supplied game.

65 Eatable means nice to eat, whereas edible means safe to eat.

its so-called "expiry date"[66] has lapsed. This is an incredible waste, environmentally unsustainable, and certainly a financial burden. I often wonder if companies are abusing this originally well-meaning concept to increase their sales. I personally trust my senses much more than any such overcautious approach. Let's look at a few simple examples.

Yoghurt and milk sold in shops is always pasteurised (treated briefly at very high temperature to kill bacteria) and sealed in an anoxic (oxygen-free, usually nitrogen) atmosphere. Milk can either be designated long life (UHT, extended pasteurisation at very high temperatures) or fresh. In the latter case, it's given 5–6 days' life which is erring heavily on the side of caution. Even after 7 or 8 days in the fridge, the milk is generally still safe, unless it was contaminated during packaging. Under normal conditions, the worst that can happen to it is that it should go slightly sour, which is not dangerous at all. Even if it curdles, it's still edible, although not as pleasant. Sour—slightly acidic—milk products (e.g., kefir) are available and apparently aid the immune system. Curdled milk is the next stage, when these acidic agents bind the proteins together with the water to make frothy, solid masses, usually floating in the milk. Yoghurt, cheese, and kefir are the next stage, made with special bacteria. All of these stages are perfectly edible, although not always palatable.

Oxidation is another threat, but milk cartons are multi-layered with an intermediate aluminium barrier and plastic inner layer which stops all oxygen permeation. Yoghurt pots are made of thick polypropylene[67] and the sealed lid is aluminium, both of which are practically impervious to oxygen. Yoghurts are given about 2–3 weeks expiry dates, but if there is no obvious mould, they can be eaten or used well after that date. I regularly do myself. Vacuum packed processed meats and cheeses are covered in polypropylene or some other oxygen barrier membrane and given a few weeks' life, although they can easily last months, especially in the fridge. And so on, and so forth.

The "best by date" is one step lower in recommendation and generally means that the vegetable or fruit or whatever may have lost some of its expected nutrient content, usually be oxidation or reactions with atmospheric humidity or exposure to sunlight. In reality, only a few nutrients are seriously affected by air over a few days or even a couple of weeks, even if the vegetable or fruit looks wilted. Surface proteins, trace elements, and sugars

[66] In most cases, "expiry dates" are recommendations, not restrictions, but you wouldn't know it by listening to so much nonsense "advice" online and offline.

[67] For a brief introduction to various materials, see later.

can be affected by oxidation,[68] and the amounts of certain water-soluble vitamins (mainly A and some of the B complex) will be reduced by the presence of humidity. UV exposure (sunlight) also affects the same vitamins,[69] but also dries foods and initiates some dissociation of protein polymers.

All in all, my opinion is that the expiry date concept is overused and misleading and buyers should use their own judgement. I think that having only production dates imprinted on every product and smelling the food should be sufficient to determine whether something is eatable or even edible, as long as people use their common sense. If in doubt, don't throw it away, but think sustainably and ask for advice. It'll save you money as well. I personally grew up without expiry date information and still rely on my common sense to determine if something is eatable (or edible) or not. If a vegetable or fruit looks right, smells as it should naturally and there is no visible contamination or infection, then it is most probably eatable no matter what the "expiry date" says.

Of course, this is not strictly true for raw meat, poultry, fish, and similar uncooked proteins. In such cases, a natural smell should be aided by making sure it is heated right through to a temperature of about 65 °C. This is the only way to be sure that the vast majority of bacteria, parasites, and moulds that the food may contain have been eliminated.

Finally, a word about tinned foods. Few food items are nowadays sold in tins, but some people still prefer them. It is a huge change from just 30–40 years ago when a supermarket used to have aisles full of tinned foods. The main worry about a tinned food is the presence of the bacterium Clostridium botulinum which causes botulism (very rarely now) and the tell-tale sign is a badly swollen tin which indicates that the food has reacted and produced a large amount of gases inside. This may be caused by the tin having been crumpled, cracking the internal protective membrane. Get rid of it without opening.

---

68 Oxidation of foods by ozone (O_3) is much quicker than with ordinary oxygen (O_2) so it's an excellent antibacterial and antifungal agent. It is also a pollutant generated by older types of electric motors and everywhere where there is electric arcing and very intense sun light.

69 The high energy of UV photons can easily break up some of the vitamins' weak bonds. It can also affect proteins and even carbohydrates.

When Freezing Goes Wrong

Of course, we can preserve foods in the fridge for a few days and in the freezer for weeks. The problem is that this doesn't work for everything, as the conversion of water to ice results in damage to certain fresh foods. Let's consider some of them.

Nearly all slow-cooked foods (soups, stews, casseroles, etc.) can be safely frozen. The proteins have been denatured by cooking, the vegetable cell walls have dissolved, and the food sauce is a smooth mixture. Meat products and fish can also be frozen safely for weeks on end with minimal effect on structure and taste. The secret is to freeze all such foods as soon as possible before natural dissociation and decomposition reactions have started breaking down the proteins. This is especially true for fish whose fat is particularly prone to decomposition.

However, many vegetables and fruit cannot be frozen. Freezing water-heavy vegetables such as lettuce, tomatoes, and cucumbers, as well as most fruit, destroys their cell structure, because of the expansion which occurs when water freezes to ice. This expansion happens because the atomic structure of ice has many open spaces between the molecules as compared with liquid water. Recall our earlier discussion on the shape of the water molecule H_2O. The H–O–H angle is about 105°, which means that the molecules cannot fit comfortably next to one another. When the water temperature drops, their vibrations reduce and at about 0 °C, the molecules can no longer move freely, but must find a stable, rigid position. The result is a hexagonal atomic structure, but because of the unusual H–O–H angle, the hydrogen bonds between H and O are stronger than the bonds between the molecules, so the structure is slightly distorted (see figure) and more open than would have been the case otherwise. The result is that ice has about 9% lower density than liquid water and expands by that much when it forms. One consequence of this is that ice floats in water.[70] Another is that everything that contains water will swell when placed in the freezer, and indeed, not just swell, because the expanded ice will break cell walls and thus cause damage to vegetables and fruit.

[70] For any iceberg, only about 1/11 of its mass is visible above the surface. The rest is underwater.

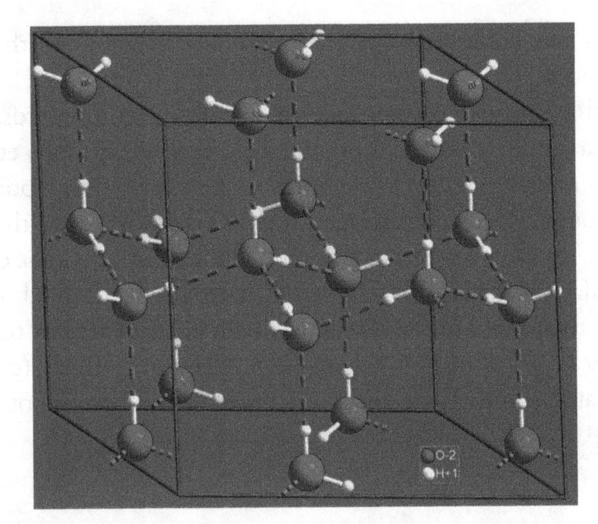

The atomic structure of ice

Not all water turns to ice at 0 °C. In order to freeze, water must at least contain some particles, for example dust or food particles, which act as nucleation sites for ice to start forming around them. This is called heterogeneous nucleation and has a much lower activation energy (minimum energy to start nucleation) than homogeneous nucleation (i.e., without particles) under identical conditions. If water is particularly clean (e.g., distilled), nucleation of ice crystals is delayed until the temperature drops to a much lower level than 0 °C (even down to − 30 °C in some cases) and the water then becomes "supercooled". Try an experiment: place very clear distilled water (or deionized) in the freezer and do not shake it as you remove it from the freezer. If you now touch it or shake it slightly, it will suddenly freeze in less than a second. Since condensed water on the underside of a lid is essentially distilled, it freezes at a much lower temperature than 0 °C and may remain liquid allowing mould to grow even in the freezer.[71] Before you freeze any cooked food, make sure you remove any condensed water from the underside of the lid.

But ordinary water can also remain liquid below 0 °C if it contains in solution certain salts such as common salt (NaCl) or $CaCl_2$.[72] In this case the mechanism for remaining liquid is different. The water molecules break up the salts and the ions float around disturbing the balance between the water molecules leaving and joining the ice at the interface. The more ions are swimming around in the water, the more they block liquid water molecules

[71] Moulds and slimes are known to grow happily even in Antarctica.

[72] Both are generally used to defrost ice on roads and other surfaces.

from entering and solidifying, and the freezing temperature of the solution is therefore reduced.

Some food plants also contain natural anti-freeze compounds. Vegetables in the brassica family (cabbage, kale, broccoli, Brussels sprouts, etc.) and also carrots contain many calcium ions, which depress the freezing point of water. They also produce special antifreeze proteins which obstruct the formation of ice crystals, another way to stay liquid. The net effect is that cabbage and kale and similar vegetables can withstand severe frost and much lower fridge temperatures than most other vegetables, although not freezer temperatures. In fact, many of them taste much sweeter after strong frosts (even Brussels sprouts), because some of the antifreeze proteins can convert other proteins to sugars.

Tastebuds Never Lie—Or Do They?

Is taste due to chemistry or physics? I think a bit of both. Taste is essentially the way our brain perceives an excitation reaction that occurs on thousands of microsensors (receptors) on the tongue and at the back and on the roof of the mouth when particular molecules touch it. Let's see how it works in some detail.

Neuron whispering

Neurons don't actually touch one another in order to deliver their message. Electrical signals, called action potentials, travelling along the neuron "jump" across very small gaps called "synapses" between tendrils and adjacent neurons. By connecting many neurons via synapses, a signal travels from the brain to the periphery and vice versa.

The jump can be either electrical (here the synapses almost touch via special proteins and the signal is transmitted almost instantaneously) or chemical, where neurotransmitter molecules "carry" the signal across a gap, which gives a slight delay. By repeating the signal, the synapses can be "strengthened" by increasing the number of neurotransmitters.

It's interesting to speculate why synapses did not all evolve to be instantaneous (electrical). The short delay in chemical synapses must offer some adaptive or protective benefit.

Food consists of various molecules, not all of which excite the taste sensors, while some excite more than one type of sensor. As you probably know, there are five tastes: sweet, bitter, sour, salty, and umami (or savoury). The taste we feel is sent to the brain by specialised neurons, where it is appropriately "translated" by the brain according to the excitation potential signature. When a molecule of sugar (or glucose or fructose, etc.) is dissolved in saliva, it reacts with a sweetness receptor which produces a small voltage signal. The more molecules react with it, the more voltage is produced, giving a measure of the amount of sugar. This is then transmitted along the neuron to the brain, where it is processed by the brain neurons and understood as something sweet. The more voltage reaches the brain, the sweeter the food is perceived to be.

Interestingly, we now know that sweet and bitter tastes are perceived by specific parts of the brain (i.e., they are hard-wired in the brain), but the other three tastes appear to be processed by a mixture of sensors in different brain areas, we don't know why yet. We also don't know how the brain actually learns to identify tastes or smells. My own understanding is that it has something to do with the fact that each sequence or chain of synapses (connecting points between neurons) produces a specific electromagnetic "signature" which inductively couples with (or "senses") some permanent sequence of synapses which formed previously when the same taste was first experienced and "saved" in the brain. Comparing the two sequences brings on a recall by association and we identify the taste or the mixtures as similar to the one experienced before. For example, if we encounter a very pleasant taste, the brain creates a chain of synapses with a specific electric field "signature". If the taste is strong enough (or is repeated frequently enough, e.g., licking a chocolate ice cream), these synapses strengthen and become permanent, persisting for years in many cases and now associated in the brain with a name or some other characteristic (also saved by another sequence of synapses). If now we happen to encounter the same food again, the brain will create a similar electric field signature which will inductively stimulate (or resonate with) the older one, and suddenly we have recall or a sense of déjà-vu. Anyway this is all speculative but it may explain all sorts of memory recall.

Certain unusual molecules are perceived as mixtures of different tastes and are sometimes easy to distinguish, sometimes not. Bitter orange marmalade contains molecules which excite both the sweetness and the bitterness sensors, giving this unusual bitter–sweet flavour. A sweet-and-sour sauce excites other combinations of sensors. The brain is actually sensitive enough to distinguish many different tastes in parallel—exactly what we aim for when we prepare an intricate stew or sauce.

Another interesting situation arises when a receptor is over-excited by too many molecules, e.g., by a very sweet food. The signal is saturated (too much voltage is produced), causing a negative perception in the brain which we sometimes call a "cloying" taste.

Colourful Food and Colour Surprises

All materials reflect incident light whether they are solids, liquids, or gases, but how is colour produced? And how come so many ingredients change colour during cooking? For example, red meat very quickly becomes grey when boiled and beautiful green herbs soon turn black.

The answer is related to both the material atomic structure and our own brain's perception of what it sees, all driven by physical phenomena. Let's look first at reflected colour.

As I mentioned before, sunlight that reaches the surface of the Earth includes most wavelengths around the visible region, from IR, through the visible part and UV-A to most of the UV-B region. There are also some microwaves. All of the visible waves are perceived by our brain (though our eyes) as different colours from deep red (in the near IR region, of wavelength about 750 nm) all the way to deep violet (in the UV-A region, of wavelength about 400 nm). But how do some foods (fruit, vegetables, meat, and others) appear coloured or even iridescent in the first place?

It all has to do with which of these waves are absorbed by the food when sunlight strikes it. The first thing to note is that most foods have a cell structure with the cells separated by very thin membranes. The cells are mostly water (containing various compounds and a few pigments), which makes them appear almost transparent when viewed by microscope under transmitted light. When a material is illuminated by sunlight, some light penetrates deep into the cells and encounters pigments which absorb specific wavelengths depending on the atomic structure and elements the pigments contain and their electron energy levels. All other wavelengths are simply reflected away, so the compound appears to be the colour of the reflected photons alone. If nothing is absorbed, the food appears white. If all visible wavelengths are absorbed, the food appears black.

Foods are coloured mainly because of five families of chemical compounds: blood haemoglobin (absorbs all wavelengths except the red colour, as in meat, fish, etc.), carotenoids (yellow, red, and orange in carrots, oranges, tomatoes, etc.), chlorophyll (green in lettuce, parsley, broccoli, etc.), flavonoids (red, blue, and purple, in raspberries, etc.), and betalains (red, yellow, and purple, in beetroots, etc.). Many foods contain combinations of these compounds, so the resulting colour we actually see may be a mixture of these non-absorbed colours, as in mauve or purple aubergines, for instance.

Certain fish can also appear beautifully coloured and many are iridescent by reflection of specific wavelengths and by multiple reflections and prismatic splitting through semi-transparent scales and skin cells. In fact, changing our viewing direction gives different reflected colours in this case.

Naturally, if the incident light contains only specific wavelengths (for example certain artificial lights), the resulting colour we see may be very different. Plants illuminated in red–blue light (in many modern hot houses) appear black because there is no green wavelength to be reflected back.[73] Finally, when the ambient light is very weak, there is hardly enough to be reflected back and most foods (or any other material) appear dark grey or black.

Now, during cooking, two things may happen. If the temperature is very high (as occurs in frying or baking), some of the pigment compounds may dissociate, and the element that gives it colour (e.g., iron oxide in haemoglobin) may change atomic structure. In the case of haemoglobin in meats, after a few minutes cooking covered in a pot, it gets deoxygenated (loses some of the oxygen) and the iron compound changes from ferric oxide (Fe_2O_3), which is red, to ferrous oxide (FeO), which is black, as we discussed earlier. That's why meat looks grey–black after a few minutes in a pot with water. However, on hot coals or under the grill, there is enough atmospheric oxygen for the meat to retain or strengthen its reddish colour longer.

Some compounds retain their colour even after cooking. For example aubergines, carrots, and tomatoes hardly change their colour during cooking, but they do become darker as they oxidise, because the slight change in their structure means they absorb more wavelengths. This also happens with many green herbs and lettuces, which all become nearly black when cooked.

Toast Always Lands Butter-Side Down, but Gravity has Its Good Side Too

Don't you sometimes wish that gravity didn't exist? Then we wouldn't drop things, breaking them or making a mess. And a slice of toast wouldn't land with the, heavier, buttered-side down. Well, be careful what you wish for. Gravity is critical for our everyday life. In the kitchen, we would hardly be

[73] One can say that green is a useless wavelength for the growth of plants. The same happens in some highways illuminated by monochromatic "yellow sodium arc lamps"—all cars other than yellow appear black.

able to cook and eat if it wasn't for the support and direction that gravity gives us.

As Newton first realised, gravity is the mutual attraction between any two or more bodies. Because the Earth is so much larger than us, everything looks like it is attracted to the floor. When we let go of anything, it invariably falls down, bouncing, breaking, or making a mess. We live in and are completely used to this gravitational "field". All our lives revolve around the need to accommodate and mitigate this strong force acting on us at all times. That's why the muscles that hold our back up are so strong. But gravity is not the same everywhere. Its strength depends on the planet we are on, and in particular, its mass and its diameter. On the Moon, gravity is about a sixth of what it is here on Earth, so a man would weigh only about 15 kg there. That's why the astronauts could be seen hopping around during their walkabouts on the surface of the Moon. On Mars, gravity is about a third of what it is on Earth.

On a space station revolving around the Earth, gravity is artificially arranged to be close to zero, by making the station revolve at such a high speed (about 17000 km per hour) around the Earth that the centrifugal force exactly counteracts the Earth's gravitational attraction. If it stopped revolving, it would immediately fall back to Earth. It's very easy to get used to living without gravity. In fact, astronauts say it's great fun not having to worry about dropping things and being able to move or lift huge things with no effort. But coming back to Earth can be a bit of a shock. Once I spoke to an astronaut at the European Space Agency who had come back to Earth after about six months in the International Space Station and he said his biggest shock was that in the first few days he was always dropping things because he didn't expect them to fall if he let go of them! He thought it was terribly funny, but also very disorientating that gravity would act on him all the time. He had already broken one or two cups that way.

But I digress again. Now let's get back to our kitchen and reflect on how important gravity is there. Our lives and actions are so well adapted to gravity that we don't even notice it. Think about it for a moment. If it wasn't for gravity, water would never stay in a glass and food would never sit in a pot while cooking. And boiling would be impossible. In fact, the pot itself would never just sit on the hot plate, but would start floating around. Water would just shoot out of the tap due to the pressure behind it and splash all over the place, instead of just going downwards into the glass. It wouldn't even be funny. When you turn on the gas on the stove, it would not burn upwards but would go all over the place, causing fires, because the buoyancy of fluids and gases depends on gravity. This means that a lighted gas flame, being lighter than air, prefers to go upwards only because gravity is acting. If gravity were removed, as soon as it came out of the gas pipe and there was no force acting on it, it would go downwards, left, or right unpredictably.

But let's assume that somehow we are able to hold a pot on a hot hob and attempt to cook something. Again, without gravity, the water and anything else we put in the pot will not make proper contact with the bottom of the pot, making it impossible to heat it up.

Let's take this idea further and assume we fill up the pot completely and close it tightly, so that the water and food has nowhere to go and has no choice but to boil. Now we have another problem. While the bottom of the pot gets hot, there is no actual boiling since there is no buoyancy. Bubbles form randomly and will not rise as expected. This means that the food will not be heated uniformly as happens in a gravitational field. Even trying to stir something in a pot will be very tricky, since the stirred food will not fall naturally back down, mixing with the rest, but will stay where you put it.

We've already mentioned that, without gravity, water coming out of a tap would not fall downwards but would splash all over the place. But what about swallowing? Could we still swallow normally? Well, here things are not so tricky. Thanks to our mouth's sucking action, a low-pressure is created behind our mouth, which means that food and water are sucked in normally and forced downwards into our stomach. Just don't expect to be able to pour water "down" your throat. When it reaches the stomach, it must still stay there, which it doesn't always like to do, and many astronauts report feeling nauseous or even suffer from regurgitation for the first few days in a gravity-free environment. Burping is also very difficult.

What about eating on a plate? Again, it is almost impossible without gravity. That's why most food on the space station is contained in tubes or sealed bags and either sucked out or eaten directly out of containers or tins. Food will not stay (unless stuck) on any plate, and the plate itself would float around happily, making it impossible to use your fork or knife, which are also floating around, of course. In fact, any attempt at cutting something with a knife would simply result in the food and the plate moving in the opposite direction to the cutting knife.

What about trying to clean something with soap and water? Again, without gravity it's a very tricky problem. The water would wet the plates and the pot with difficulty and the soapy water would again float around aimlessly, instead of wetting the dirty surfaces. Better to use paper plates.

In a nutshell, we should be glad we have gravity to show us where up and down are, keeping food on our plate and drink in our glass.

What Every Coffee and Tea Lover Should Know

Do you enjoy your morning coffee or tea? I'm sure you do, like me and many millions of others. It's one of the sweet pleasures that we look forward to every morning (and mid-morning, midday, afternoon, etc.). But there are a couple of interesting aspects of both coffee and tea that are worth knowing in order to enjoy them to the full and get maximum benefit from them.

Coffee and tea contain dozens of compounds that range from caffeine to polyphenols and flavonoids, which are considered to be anti-oxidants. Interestingly, if you enjoy a dark coffee or dark tea for their "fuller taste" (whatever that means), you are probably sacrificing many of these compounds because of the high-temperature treatment that is required to make them dark. In particular, dark coffee contains only a fraction of polyphenols and other compounds that are contained in lightly-roasted coffee, and the same is true for tea.

The reason is that high temperatures increase the atomic vibration of the atoms that make up these (rather sensitive) chemical compounds, leading to dissociation in many cases. This is exacerbated in the presence of very hot water when you make your brew. Tea in particular is very sensitive to such dissociation, which explains why many teas require only moderate water temperature for maximum benefit, some green teas as low as 60 °C. This essentially means that, if you want a strong pick-me-up in the morning, use lightly ground coffee or green tea, both made with moderately hot water and short brewing time. And if you just want a nice, tasty coffee or tea before bedtime and then enjoy a good night's sleep to boot, by all means have a dark coffee or tea made with nearly boiling water.

Making a cup of coffee involves a number of interesting phenomena and each method of making it has a different taste as a result. Let's consider some of the methods used to produce coffee drinks—actually coffee suspensions, since they all consist of solutions of soluble compounds mixed with suspensions of non-soluble compounds.

The simplest method is filtered coffee. Boiling water, at a temperature above 100 °C since it is heated under some pressure in the machine, is poured through the filter holding the coffee in the form of a loose, coarse powder. This coarse powder has a rather low specific surface area (about 1 m^2/g) and the water remains in contact with the coffee for a very short time, so a high temperature is necessary to enable some diffusion of the caffeine and other compounds into the water. This means that many of the most sensitive, aromatic compounds are dissociated and lost. The result is that filtered coffee acquires a slightly over-heated flavour and loses nearly its entire aroma. Not for me.

Instant coffee also compromises aroma for the sake of convenience. It is made by boiling, distilling, and drying ground coffee, so the granules we see are porous agglomerates of nano-sized particles of very high specific surface area (over 100 m^2/g) and therefore appear to dissolve completely in water. In reality it is a solution plus a suspension of nano-sized coffee particles. The taste can be very strong since it is made by distillation, but it has almost no coffee aroma. Using water at 90 °C recovers some of the aroma but only for very short time and it has a very limited taste. No again, thank you.

Espresso coffee is made very differently. The coffee is ground very finely (below 10 μm) so its specific surface area is moderately high (over 10 m^2/g) and packed quite tightly in the machine. At the same time the water is heated to only about 90 °C, but forced under high pressure through the packed powder. This gives limited flow and a limited amount of liquid. The result is a highly aromatic and very dense coffee, but a very small amount of it. Much better. Macchiato is similar and even more intense with added drops of milk.

Nanoparticles old and new

Micro and nanoparticles (generally speaking anything smaller than about 5 μm) are everywhere around us in the form of fine dust, soot, pollen, shed skin cells, etc., and most of it is non-degradable and indigestible. When we breathe it in, our lungs can get inflamed, covering it with mucus and trying to expel it before it causes damage. If we swallow it, some of it gets dissolved and digested, but most goes right though us.

Plastic micro and nanoparticles present a particular problem, as many polymer chains are extremely stable. The ones that are bio-based (cellulose, polylactic acid, etc.) are probably broken down and digested to some extent, but most of the synthetic plastics are probably not and pass right through. However, when they are particularly small, smaller than about 30 nm, they can diffuse through the intestine walls and eventually reach various organs. The jury is still out whether such, otherwise inert, materials can cause serious health problems, although animal studies indicate that they can.

Greek coffee is quite different again. The coffee is ground very finely (below 2 μm), so its specific surface area is quite high (over 50 m^2/g) and a much smaller amount is used than in espresso. It is added directly into a small boiling pot with water, stirred well and allowed to heat up slowly and boil only momentarily so that it preserves most of its aroma. The Turkish method uses the same type of coffee but the coffee drink is allowed to boil for a while, so it has less aroma but a stronger taste, since more compounds have time to dissolve or diffuse into the water. Both these methods result in a rather dense solution with a suspension of nano-sized coffee particles. My favourite, of course.

Lately, there have been a number of articles presenting research findings that all hot coffee and tea (and other hot beverages) sold in plastic-coated containers contain billions of plastic nanoparticles, probably leached out from the inner surface of the cups. While this is certainly a cause for some concern, it is also true that all such beverages contain billions of their own coffee and tea nanoparticles, mostly in the form of agglomerates of certain compounds. In fact, all filter coffee and tea beverages (and of course all Greek- and Turkish-style coffees) contain not only huge numbers of nanoparticles but lots of particles at the micrometre level as well, all of which we ingest. I know people that actually love to eat those fine sediments and would never consider discarding them.

Storm in a Teacup

When I make a cup of tea or coffee, it is fun to observe the various physical phenomena that go with it. We've talked about the diffusion of tea or coffee compounds into the water, but there are other interesting physical phenomena too.

First of all let's look at the shape of the tea surface as you stir. When you stir it, its surface gradually climbs upwards around the cup circumference (by the centrifugal force of the rotation), but forms a hole at the centre. The faster you stir, the higher it climbs and the lower the centre hole gets, right down to the bottom of the cup. You have created a vortex. If you look at this vortex shape from the side (an electrical stirrer in a transparent glass cup of tea with milk shows it very well), it looks like a smooth curve. The shape of the vortex is a paraboloid, one of the most commonly encountered types of curve in nature. The exact shape is the result of the need for the momentum of the tea in the middle to keep up with that at the edge. It's the same shape of curve as many flower petals (lilies, etc.), the curve of the rainbow, and the curve of the vortices made by a body moving fast through water. It is also the shape of the distortion that masses make in space according to Einstein's general theory of relativity (gravitation theory). Not bad for a cup of tea.

Let's look at what milk does in hot tea. Milk is also both a solution of water-soluble nutrients and a suspension containing solids and, in full-fat milk, various lipid molecules. If you pour skimmed milk in tea, it mixes very easily, but you can still see the vortices forming as the milk goes in. Full-fat milk is more interesting. Vortices form very quickly and remain stable for some time while the lipid molecules gradually melt but stay separate from the water. Eventually, the lipids form islands on the surface as their density is lower. Adding just a drop of brandy or other alcohol (or an acidic substance like lemon) will quickly help to bond the lipids with the water and smooth out the ugly islands. Or just use skimmed milk.

Talking about milk in tea, there is a lot of discussion about which is correct or preferable: to pour the milk before or after the tea? I think it is a case of personal preference, but I also think that pouring very hot tea (freshly brewed in a teapot) onto a small amount of milk will scald it and dissociate the molecules in it, but not the lipids. Pouring milk onto tea is less damaging since the tea has already had a chance to cool down. You can get the same result by pouring tea at less than 90 °C onto milk. I personally only drink tea without milk. But there's no accounting for taste, of course. All the above are of course also valid for coffee with milk.

Now let's leave the milk out for a moment, stir, and observe the tea leaves. Can you see them making a "boiling" cyclic ("helical") movement while you are stirring? They move to the outer periphery, sink along the outer edge, move a little towards the centre and then rise again in the middle.[74] What is happening is that the leaves get thrown sideways by centrifugal force, meet some friction at the periphery, slow down, and sink to the bottom. They then move inwards, hit the rising part of the liquid, and rise to the top to complete the cycle. But why do they move inwards at the bottom?

Let's look a bit more closely. In the same cup of tea without milk, observe the movement of tea leaves at the bottom of the cup as you stir a little and then stop. Initially, the leaves move to the periphery, but as soon as you stop, they drop down and start congregating at the centre. Fascinating and entertaining. But why?

This is a rare demonstration of the centripetal force pointing towards the centre of the rotation. Generally, the centripetal and centrifugal forces are always balanced (as when you swing a ball on a taut string around your head). In the case here, when the tea leaves reach the bottom and you stop stirring, they encounter some friction, but the liquid still rotates, so the centripetal force on the leaves is now greater than the centrifugal force and they slowly move to the centre. If you don't stop stirring, the tea leaves meet the faster-rotating vortex at the centre and get pulled up, away from the bottom surface, following the parabola to complete the cycle. Interestingly, the same phenomenon occurs in the sharp curves of rivers, where stones tend

[74] Use green tea leaves or add a few drops of lemon to lighten up the tea (by bonding with some of the tea molecules) in order to see their motion more clearly.

to congregate at the inner bank of the river, not the periphery. A slightly counter-intuitive fact, but quite true.

Let's now add some milk and observe the surface of the tea as it cools down. If you use normal milk, you'll see some floating "scum". This is the bane of all tea drinkers and occurs even in black tea. It's different from soup scum and what happens is this. All tea leaves (in fact, nearly all leaves) are covered by various insoluble waxy compounds which, when you pour in hot water, are released and float to the surface. In addition, in areas of hard water, various carbonate minerals (especially calcium carbonate) in the water bond preferentially with the waxy substances, exacerbating the scum problem. Even worse, when you add non-skim milk to hot tea, on cooling, tiny fat globules tend to separate out and coagulate with any waxy or salt substances, making your tea look terrible when it cools down. The best solution is to drink your tea hot.

5

Tools of the Trade: Appliances, Materials, and Trusted Kitchen Helpers

"Happiness is a small house with a big kitchen."
- Alfred Hitchcock

G. Vekinis, *Physics in the Kitchen*, Copernicus Books,
https://doi.org/10.1007/978-3-031-34407-7_5

I hope by now you are convinced that physics is everywhere you look when you cook. But what about the rest of the kitchen? How much physics is around us when we prepare and cook a meal? Let's have a closer look.

As I said at the beginning, physics is everywhere in the kitchen. All our appliances and materials and "labour-saving" devices are built and operate strictly according to the laws of physics. From the ubiquitous stove to the lowly kitchen timer, everything that engineers design and build obeys specific physical principles. Let's visit some of them and see for ourselves.

Gas or Electricity?

This is a major dilemma all over the world at present. The majority of kitchens globally are using gas hobs for cooking (as well as heating water), a method of heating which is terribly unfortunate from an environmental point of view. Originally, it was of course a foregone conclusion that natural gas should be used since it was (and in many countries still is) the cheapest and most convenient way to cook and heat water. However, our belated realisation that our unchecked use of fossil fuels has resulted in serious changes in the global climate means we must move away from all fossil fuels as soon as possible. And this includes the beloved gas hob.

Wood-burning climate damage

There seems to be a completely misguided idea that using wood-burning stoves does not affect the climate, since wood is sustainable and does not release fossil-stored carbon dioxide. The thinking goes that since this carbon is recycled it does not contribute to global warming.

But this is just wishful thinking as it does not take into account the direct environmental pollution as well as the indirect climatic effect of that pollution.

Burning of wood (and coal) in a stove is never complete combustion (that would require temperatures above 1000 °C), but produces a plethora of dangerous soot and silicate particles, as well as "aromatic" hydrocarbons and carbon monoxide, most of which are toxic and carcinogenic. Taking all that into account, it's easy to conclude that wood-burning stoves are a clear liability for our health, the environment, and the climate.

Gas provides an immediate means of heating a pot. You can control the temperature and cooking process almost instantaneously since combustion takes place exactly underneath the pot. Aluminium, copper, and thin steel pots and pans are ideal as they have very low heat capacities and heat up very quickly. They can also be used on any gas fire, even if warped or bent. Using such lightweight pots and pans, you can sear a steak and prepare a sauce quickly and almost effortlessly. And gas used to cost next to nothing. Gas is also available in bottles (RPG gas) so it's an ideal fuel for homes in remote locations or for camping.

Before we discuss the pros and cons of gas ranges as compared to electrical ones, let's have a quick look at the physics involved.

Most cooking and heating gas consists mainly of natural gas (about 95% methane, which is odourless) with the addition of a little butane, which smells very strongly to warn us of any leaks. After it goes through a pressure controller, it is fed into the gas hob via copper pipes. The main part of the hob is the mixer and burner crown, where the gas is mixed with air at a gas:air ratio of approximately 1:10, and the mixture is vented out through the holes in the crown, where it is ignited in a self-sustaining exothermic reaction.

Gas bottles (large and small) are highly pressurised so most of the gas is in the liquid state with gas above it in equilibrium.[1] You can easily hear it sloshing around. All gases become liquid if pressurised enough, because the molecules are brought close enough by the pressure that their attraction overcomes their kinetic energy and they become liquid. However, they transform back to gas as soon as the pressure drops, when we turn on the valve. The pressure in a gas bottle at room temperature (23 °C) is about 2 bar,[2] but it increases a lot on a hot day or if it is exposed to the sun, so it has to be kept in the shade to avoid an explosion.[3] It also decreases a lot in cold environments (e.g., outside in winter), sometimes so much that the gas supply all but stops because the pressure in the bottle is not enough to push the gas out.

Cooking gas bottles do not contain natural gas but liquid petroleum gas (LPG), which is produced from fossil oil and consists of a mixture mainly of propane (70%) and butane (about 30%). Although the combustion temperature of both LPG and natural gas is about the same (around 2000 °C), LPG provides more than double the heat energy per unit volume of gas. The gas from the bottle goes first into a pressure reducer (controller) before it is fed into the kitchen hob. Small camping gas bottles contain mainly butane and have no pressure controller. The gas is fed directly into the small mixer/burner crown after sucking in air by the Bernoulli effect. This is done through the two little holes just under the crown. The gas moves at high velocity inside so the surrounding pressure drops and this is what sucks the air in.

---

[1] Gas molecules continuously transform to liquid and vice versa because of the randomness of their energy content.

[2] 1 bar is equal to atmospheric pressure at sea level, about 100 kPa or 0.1 MPa.

[3] If the gas bottle is in the sun, its temperature can increase to above 70 °C and the internal pressure can reach over 12 bar. Highly dangerous.

Gas has been the mainstay of kitchen hobs (private and professional) for decades because of its low cost and speed of use. But things are now changing since the price has increased and with the realisation that gas combustion produces significant amounts of pollution as well as carbon dioxide (the main greenhouse gas responsible for climatic disruption), even when it's well maintained. In fact all fuel gases have other disadvantages that are not obvious. First of all, methane is completely odourless (to warn us of leaks, they add a little smelly butane) but toxic and deadly in sufficient quantities. Moreover, all gases are highly explosive in closed areas. If the hob is not well adjusted or is not in pristine condition, the gas does not combust completely as the gas:air ratio increases beyond the optimum, and thus oxidation is not complete. This results in the production of carbon monoxide and soot, plus cyclic aromatic hydrocarbons and nitric and nitrous oxides, all of them highly toxic or carcinogenic. Recent tests have also shown conclusively that even well-adjusted gas hobs produce large amounts of fine particulate pollutants, as bad as living right next to a major gridlocked road.

Soot at home

Soot is well known as a product of wood and wax flames, but they are not the only sources. All organic fuel flames—gas, oil, petrol, diesel—produce many different types of soot and, it is now recognised, they are all dangerous.

Some flames produce visible soot, but it is the invisible soot that is most concerning. In general, the smaller the soot particles, the more dangerous they are. It is now well established that the most dangerous soot pollutants are those smaller than 2.5 μm, usually generated by internal combustion engines, especially diesel engines. Unfortunately, gas flames produce a large quantity of these tiny particles too, many of them very stable in the form of, otherwise beautiful, spherical fullerene molecules. Better keep the extractor fan on.

Finally, most of the heat from a gas hob flame goes to heat up the room (up to 60% of the total) rather than the pot and it generates a large amount of water vapour as well,[4] certainly not welcome on a humid summer's day. Maybe it's time to seriously reconsider using gas ranges in the kitchen and change to electric.

Electrical ranges on the other hand do not have these disadvantages, but they are slower to heat up and require a flat and level hob and flat and level pot and pan bases to avoid energy loss due to bad thermal conduction. But they have no direct emissions in the kitchen from combustion at the hob. Of course, such an advantage goes out the window if the electricity is produced

[4] Under optimum conditions $CH_4 + 2O_2 \rightarrow 2H_2O + CO_2 + \text{heat}$.

in fossil-fuel power stations in the first place. It's just a case of displacing the emission source. At present the majority of electricity world-wide is still produced by such polluting means, although renewable (or nuclear) sources are gradually replacing them. Not a moment too soon.

There are two types of electrically heated ranges. First we have the traditional one, where the resistive "elements" (coiled wire) are buried inside a metallic plate which does not show any changes when heated. Secondly, we have the black glass–ceramic topped range, where the resistive elements are coils just underneath the ceramic top and glow red when heated. Both work by exactly the same principle. By supplying electrical energy to a high-resistance wire (usually nickel–chromium or molybdenum-based), we increase its temperature as the electrons interact energetically with the metallic atomic lattice (Joule heating). In the case of the hot metal plate, the heat is distributed evenly throughout the plate by conduction and this heats the pot also by conduction. In the case of the ceramic hob, there is no contact between the coil and the ceramic so some heat is lost underneath, but it still heats the area above it. Because such a glass–ceramic has very low transverse thermal conductivity, the area directly above the coil is heated preferentially.

Both these types of ranges use the same principle but there are some interesting differences. The metal plate hobs are strong and almost impossible to break but radiate a lot of their heat. But because you cannot see the heat, it isn't difficult to forget such hot plates switched on without a pot on (to absorb the excess energy) so they will quickly overheat and eventually warp. At this point it either becomes useless or its efficiency is significantly reduced, although not completely so if it is only slightly warped. Since its surface is matt black, its emissivity is nearly 1.0, which means it is still able to radiate a good deal of heat onto the bottom of the pot, even if the latter does not sit properly flat anymore. It's the same if the pot is slightly warped. But it's better to get a new pot or hob anyway.

On the other hand, a ceramic top cannot warp but it can break if something heavy is dropped on it. However, because it is extremely shiny and flat, its emissivity is very low (even if it's black it's only about 0.2) and hardly radiates heat. If a pot does not sit perfectly flat on it, it will hardly heat up. When switched on, one can easily see the red hot coil. But when switched off, because it doesn't radiate much heat even when very hot to the touch, there are red indicator lights on such tops to warn you not to touch the hot areas after switching off.

Induced Magic

In addition to the traditional gas or electric hobs, there are also inductive heating hobs which are claimed to be very energy-saving and convenient. How do they work and is the hype justified?

As the name implies, inductive heating is all about an alternating magnetic field inducing a current in a conductor. This was first discovered by Michael Faraday in 1831. However, in this case we do not use a wire resistance to create heat, but the metal of the cooking pot is heated directly. Heating happens because a high-frequency alternating magnetic field, i.e., with flip-flopping direction, produced by an electromagnetic coil under a flat glass–ceramic top induces strong currents within the metal base of the pot or pan. These currents are called eddy currents.[5] They heat up the base of the pot just as an embedded resistive wire heats up the traditional hobs. So, induction heating is a combination of electromagnetism and ordinary Joule heating.

Can you see the dilemma here? Your copper, aluminium, and even certain types of steel pots and pans cannot be used with induction hobs because they have low electrical resistance and cannot be heated efficiently by induction. For this reason, induction pots and pans are made of metals with relatively high electrical resistivity such as carbon steels, ferritic stainless steels, and cast iron clad with ceramic enamel for corrosion protection. These types of materials all have high electrical resistance, so they heat quickly and efficiently.

Induction heating is very quick—it only takes a few seconds for the base of a pot to get very hot, so they are ideal for soups and similar liquid foods. They are also excellent for searing meat in a frying pan within a few seconds. As soon as you switch off the hob, the heating stops, although the base remains hot a little longer. Hence, because of their speed and the fact that they do not need to heat up a hot plate or the intervening ceramic top, they save electricity, as much as 10–20% in some cases. This immediate response is very useful in helping to prepare foods very quickly. On the other hand, induction hobs require a lot more current than ordinary hobs to operate and the coil needs a strong cooling fan to keep it cool and avoid damage. This means that it's often a bit noisy and your home electricity supply must be of the appropriate capacity.

Induction hobs have a few further drawbacks. Because the base is heated immediately, the sides of the pot remain cool for longer, so the food must be stirred continuously, especially stews, otherwise the food will pyrolyse easily

[5] Similar to the eddies in a fast-flowing stream.

at the bottom. To avoid this, there are temperature sensors embedded under the pot, switching the electromagnet on and off to regulate the inductive heating. Unfortunately, this sometimes creates an on–off heating problem in the pot as well. Naturally, for the induction to work best, the pot must be sitting snugly on the hob, as with traditional hobs.

The cooling fan is not the only noisy problem with induction hobs. The high frequency alternating magnetic field means that the pot vibrates slightly at that frequency (about 20 kHz, above our hearing range). Even if it sits perfectly on the surface, the vibration or "hum" is quite obvious, as the pot is being directly influenced by the magnetic field.

The March of Electrons

Once a student asked me what electrons actually do when they reach a heating element or an electronic circuit for that matter. It's another case of something being so ubiquitous that it's hardly ever explained in detail, even in physics courses.

Earlier we mentioned that electrons move through a wire very slowly but, because they push against each other like little soldiers in a tight line (or tightly packed dominos), their electrical effect is almost instantaneous. The continuous supply of electrons is of course called the electric current (measured in amperes). It's exactly like water gushing out of a tap as soon as you turn it on, or like fans squashed against the gates of a stadium to enter for a game. In fact, electrons are always under "stress," just as water in a pipe is always under pressure. They are being pushed forward by the electric potential behind them (from the battery or a power supply), and if the wire is not connected anywhere, or if it is cut, for example, by a switch, they cannot move forward.[6] So when we turn on the switch and close the electric circuit, the electrons surge forward into the circuit, where they give their energy in the electronic circuit or deposit their energy as heat.

We saw previously that a current of electrons heats up a resistive element in a hob by electrical interference with the atomic lattice, and that there is always conservation of particles—what goes in must come out. So, where do the electrons go when they have used up their energy or done their work in a circuit? Well, some of them attach themselves where they gave up their energy, some follow the circuit and go back to the source (that's why we have at least two wires in any connection, the "live" and the "neutral"), and some of them

[6] Actually, there is always a tiny leakage of electrons into the atmosphere at the end of any wire. This is exacerbated by humidity. Only in a vacuum is there no such leakage.

leave the circuit completely by leaking into the atmosphere or leaking into the ground via the body of the machine. This can happen either directly or via a third wire (the "earth" wire) in the plug. In fact, electrons would love to go straight to earth, the lowest energy level, and not be bothered with doing work in a circuit. In many countries, the neutral wire from the source is often connected to the earth after it reaches your kitchen. This way, most electrons, after depositing their energy, are returned via the neutral wire directly into the earth and relatively little current actually returns to the original source. By the way, this type of connectivity plays havoc with some delicate and precise measurements in physics labs, but that's another story.

We might well ask then, why do we even need a neutral wire? Wouldn't the live wire and an earth connection be enough? In fact, it would not, because the electrons would then choose the easy way out and go directly into the ground (a "short circuit to earth") without participating and giving their energy to the electric or electronic circuit. In any case, earth connections on machines are necessary if they have a metal casing, like a fridge, a toaster, or a stove. The idea is to send the current directly to earth if a broken live wire should happen to touch the case, and hence avoid anyone receiving an electric shock.

Talking of receiving a shock, have you ever been stung by static electricity? By a small (or not so small) arc between your fingers (or elbow or any bare skin) and an earthed metal? It's rare for it to happen in the kitchen because of its relatively high humidity, but not impossible. It generally happens during very dry periods, when trillions of electrons accumulate on us by various processes (shuffling shoes on carpets, for example) and are desperately looking for ways to return to earth. As soon as we get close to a metal structure (a metallic lift button is a "popular" place for this), they jump across the gap, using the air as a conduit and giving us a stinging taste of plasma. It can be a surprisingly large amount of potential difference (voltage) by the way, about 30,000 V per centimetre of arc. This can't burn us though, because the amount of current is very small, a few microamperes (μA) at most, but it can still cause damage to some sensitive electronic equipment. Technicians always connect themselves to earth when they work on a computer or a smart phone to avoid damaging sensitive electronics.

Static electricity builds up all the time, but we never see an arc during humid periods. That's because electrons readily jump onto minute water droplets in the air before they have a chance to build their numbers up on our skin, and also because humid air is a better electrical conductor than dry air and forms a continuous connection with the ground (the earth).

Let's now see what electrons do in some electronic devices we use in the kitchen: one has a small cavity that produces penetrating "light", another uses a little hammer to produce a spark to light the gas hob, yet another tells us the time, a fourth reminds us to switch off the oven, others to communicate, etc. Electricity is the central energy source for most everything we use.

Microwaves—The Quiet Revolution

As we discussed at the beginning, microwave radiation (MW) is just another form of light, but invisible to us. It is situated before radio waves in the electromagnetic spectrum, at slightly higher energy. This light consists of much lower energy photons compared to those of infrared (IR) or visible light. Although the energy of their individual photons is low, they can penetrate up to 15 mm into some dry materials (foods, plastics, glass, ceramics, etc.) whereas visible light cannot. However, they cannot penetrate any metallic material, no matter how thin it is. This is because metallic materials are electrically conductive and do not allow penetration.

As we discussed before, MW light is ideal for heating up foods that contain water or some types of butter that can absorb MW energy. Wet foods such as ready meals, sausages in a bag, leftovers, soups, or a glass of milk are ideal candidates for heating by MW energy, but it is not very good for broiling a steak or browning a cake. But all in all, they have made life very comfortable for many people since the discovery of their heating effect in the 1940s.

There are a lot of misconceptions about microwave heating, which is unfortunate because they can save us loads of electricity. Modern microwave kitchen ovens are perfectly safe, very convenient, and very energy efficient. The energy of MW photons is about 1000 times lower than the energy of infrared (IR) photons, whose energy is about 100 times lower than visible light photons and another 1000 or so times lower than UV light. You can stand in front of a home heater (which heats you by irradiating you with IR light) for hours and nothing will happen to you, except get hot. But if you lie more than a few minutes under an ultraviolet (UV) tanning lamp you have a good chance of getting burned or worse. It never ceases to amaze me that some people are perfectly happy to stand under the hot sun for hours, being irradiated by ultraviolet rays, but believe that microwaves are somehow more dangerous. It's obviously illogical since UV photons carry millions of times more energy than microwave photons. I guess it's all about believing misinformation or simple misunderstanding. I personally have been irradiated by microwaves in my research work with the only effect being a slightly warm feeling under my skin.

Before we look at the details of microwave heating in the kitchen, let's see how microwaves are produced, without going into details. MW light is produced similarly to radio waves, by feeding electrons into a special cavity called the "magnetron," where they create EM vibrations with a frequency between 300 MHz (millions of Hertz) and about 100 GHz (billions of Hertz). These are fed onto a small antenna (you might have seen it—it looks like a metal-capped, pink ceramic finger) and then through the "waveguide" and a plastic or mica "window" into the oven chamber where they irradiate your food.

Now let's see how MW heating works. Just like any light, MWs are electromagnetic waves (also photons[7]), but they oscillate with very high frequency, about 2.45 GHz for home ovens.

MW oscillations couple with the polar water molecules electrostatically. These molecules then shake and vibrate at the same frequency in every direction. This sets up a vast number of electrostatic interactions between water molecules, all of which vibrate violently, generating energy which appears as heat in the water.

The heating effect of microwaves was discovered by accident. Apparently, during the 1940s, scientists in the UK were developing radar (which works by sending and receiving beams of microwave light) and some of them noticed that, when they were standing in front of one of their experimental radar antennas, the chocolate in their pockets kept on melting, even in winter. They were being irradiated by so much microwave energy that their water- and butter-containing chocolates melted. They put two and two together and that's how the microwave oven was born.

The penetration depth of microwaves and the efficiency of heating varies according to how much water the food contains, since water preferentially absorbs the energy of the microwave photons. The more water the food contains, the less MW radiation can penetrate as the water absorbs most of their energy. This is the reason why trying to heat up dry food in the microwave is mostly a waste of time and energy, and the reason why soups heat up well at the surface but not deeper inside and need stirring. This heating mechanism works with any polar molecules such as butter and alcohol too.

As mentioned, most non-metallic containers or packaging (ceramics, plastics, paper, etc.) are invisible to microwaves, which go right through them as

[7] The duality of light (light behaves both as waves and as particles) is of course a fundamental tenet of quantum mechanics, as we discussed earlier.

though they are not there. Indeed, they cannot absorb any MW energy and do not heat up as a result. Interestingly, MWs go right through most types of glass just as visible light goes through glass. Visible light and MW is able to penetrate right though glass and clear water, as they are both amorphous materials, i.e., they do not have a regular atomic lattice. This means that their quantum mechanical energy band structure is too mixed up for any electrons to be able to absorb energy by climbing up to an allowable higher energy level.

But nearly all UV light and about 30% of the energy of IR light gets absorbed by glass, which gets hot by the energy gain. MW photon energy is too low to be absorbed by electrons, so MW photons penetrate glass without any energy loss and the glass hardly knows it. Interestingly, a few types of glass do get a little warm in the MW oven as they contain some metallic oxides (e.g., lead oxide) which, for quantum mechanical reasons, do absorb some MW radiation.

Many believe that no metal implements should be placed in the MW oven, but that's not true. Metal racks and rounded spoons can be safely placed in the microwave oven, but not sharp knives or forks because they have sharp edges, which means that the electric field will be concentrated there and cause arcing. Arcing can also occur if you put in a rough metal object or food wrapped in a metallised packaging, especially one with sharp edges which concentrate the MW energy. Finally, arcing can also occur if the inside of the oven is allowed to become very dirty with food which eventually gets carburised, forming electrically conductive sharp spots on or close to the waveguide window.

Microwave energy is not very well distributed in a microwave oven because of the relatively long wavelengths involved (about 12 cm), but also because of the way the light waves get reflected around inside the oven, and because, where the waves meet, they form "standing nodes" which create hot (and not so hot) spots within the volume of the oven. To even out the microwave energy, most ovens have a rotating plate where the food is placed during heating to ensure that it moves regularly through hot and cooler spots. To improve the heating distribution even more, it helps to place the food or liquid slightly off-centre on the rotating plate. Some ovens even have a few concave or convex structures around the inside metal surfaces which increase the number of wall reflections and help spread out the energy.

Finally, when the oven is running, the air ducts must not be obstructed as air flow is important, not just to remove any steam generated by the food, but also to protect the magnetron, which gets very hot due to energy losses when it generates MW energy and needs cooling.

Alarming Smoke

Everyone should have at least one smoke alarm at home, ideally inside or close to the kitchen. They have often proved to be life-savers, as they can detect the very beginning of a fire, even before we can see any smoke. All inflammable materials (papers, fuels, plastics, clothes, foods, etc.) emit very small particles as they heat up beyond a certain temperature, before any visible smoke or flame appears. If we are warned about them in good time, we can correct the problem before any flame develops.

Smoke detectors are of two types: ionisation and light-obstruction. The former is much more sensitive as it can detect the very small early particles I mentioned above, and is even able to warn us if we are about to overcook the steaks or sausages in the oven! The latter is activated when a beam of light is obstructed by smoke, so it only works after a good amount of smoke has already been produced.

The ionisation detector works by an ingenious method that involves a type of radioactivity (yes, radioactivity at home!), emission of alpha particles.[8] The material emitting them in a smoke alarm is a tiny amount of the isotope Americium 241 which has a half-life[9] of more than 432 years—enough to last for my lifetime and that of my great, great … grandchildren. A battery sets up an electric field in a little cavity into which the alpha particles are emitted and this creates a small, constant current. If some smoke particles appear, the alpha particles stick to them and the current changes, an event which sets off the alarm—a terribly piercing noise with an intensity of about 90 decibels that can wake up the heaviest of sleepers. I can vouch for it personally.

Half-life

The half-life of any radioactive isotope material is the time it takes for half of it to "decay" and transform to a different isotope, and thus for the intensity of its radioactivity to decrease by half.

For example, ^{60}Cobalt (with 27 protons and 33 neutrons in its nucleus), which is used in cancer therapy and for sterilising vegetables, decays to ^{60}Nickel (with 26 protons and 34 neutrons) with a half-life of about 5.27 years. This means that, after 5.27 years, from a starting 10 g of ^{60}Co there will be left 5 g each of ^{60}Co and ^{60}Ni.

^{235}Uranium is used in fission electricity-generating stations and has a half-life of over 703 million years—that's why its storage as a waste after use is such a difficult problem.

[8] They are actually nuclei of helium gas: 2 protons and 2 neutrons together.

[9] Am^{241} contains 95 protons and 146 neutrons in its nucleus.

Unless someone tries to open the small sealed cavity where the Am^{241} sits, such smoke detectors are perfectly safe, but they should not be discarded with the rubbish. The energy of alpha particles is low, much lower than that of the other two types of radioactivity: the electrons of beta (β) emission or the photons of gamma (γ) EM radiation. As a result, alphas cannot penetrate skin or even a thin sheet of paper, but they can cause internal damage if anyone should inhale or swallow the Am^{241} particles.

Interestingly, this same Am^{241} isotope is also a candidate for use in long duration radioisotope thermoelectric generators (RTG), which generate electricity by the Peltier thermoelectric effect.[10] These are sealed systems generating no waste and can last for years. Perfect for use on spaceships or on the Moon or Mars.

(Mal)odorous Extraction

Have you noticed that the extraction fan above the stove sometimes works well at removing unwanted smells and smoke, while at other times it doesn't? Assuming the device itself is working well, there are two possible reasons to explain this. You either have no open windows in the house or there is a strong wind blowing outside. It all has to do with air movements and how they affect air flowing through a pipe. Let's see what exactly happens.

An extraction fan works just like a vacuum cleaner. In order for it to extract air, the air must be able to move freely and flow through the fan. Since air is everywhere, it cannot be split or separated, since this would create a vacuum and fresh air will rush to fill the empty space. So, in order for the extractor fan to remove the smelly air, there must be an inward flow of fresh air from somewhere which will fill the place of the smelly air extracted. If the windows are all closed and there are no leaks (modern houses nowadays are so well insulated that useful leaks under the doors are rare), this is impossible and the extractor fan cannot do its job. The smelly air will just stay inside and the most the fan can do is to churn the smell and the smoke around. If you see condensation forming on the kitchen windows in winter when you are operating the extractor fan normally (and its filter is not blocked), it means that there is insufficient air flow through the extractor and you need to open a window to enable air flow. By the way, in winter it's better to open a window further from the kitchen so the cold air coming in by the fan's suction has

[10] A temperature difference between two dissimilar metals can generate an electric potential difference between them. The Seebeck effect is the opposite of the Peltier effect and is utilised in electronic thermometers ("thermocouples").

time to warm up (and freshen your home) before it gets sucked out.[11] A window too close to the fan in the kitchen will bring in cold air which will go out again quickly, cooling the air in the kitchen in the process without freshening up your home. Exactly the same flow problem can occur with extractor fans in bathrooms and cars.

Incidentally, it is worth noting that, because modern homes tend to be very well insulated and sealed (to conserve energy), a build-up of carbon dioxide can easily occur, unless there is frequent aeration. Ironically, older homes are noted for their ill-fitting doors and windows which help to ensure continuous fresh aeration.

Going back to our extractor fan in the kitchen, what about the case where a window is open but there is still no or hardly any extraction, assuming the filter is not clogged? This is an unusual situation and happens when there is very strong wind outside. The physical phenomenon behind it is the same as that which keeps airplanes in the air, causes hurricanes to suck windows out of buildings, and allows perfume atomisers to draw perfume up a thin tube to spray it around in very small droplets. It's Bernoulli's principle (we saw it earlier too when we discussed small gas bottles). It goes like this. Whenever air flows quickly past any surface (air, liquid, or solid, the effect is the same), it creates a drop in pressure close to the surface, encouraging other air to rush in to fill the space. If now the wind is blowing strongly outside an open window parallel to the building (and no other windows are open), it will not allow any air to enter the kitchen and will instead try to suck air from the inside to the outside against all the extractor's valiant efforts. In such cases, it would be much more efficient to switch off the extractor fan and open a window in the kitchen and another window or door in the opposite part of the house where the wind is less strong or blowing in the opposite direction. Bernoulli's principle will do an excellent job of removing the smell and smoke!

The same problems occur in fireplaces or wood or pellet-burning stoves. Unless there is some air flow in the house, the smoke from the fire cannot rise up the chimney and the house will fill up with smoke, especially at the beginning when the fireplace and chimney are still cold. When there is a strong wind outside, Bernoulli's principle means that there will be no suction if there is no air flow. On the other hand, if there is too much air flow, the wind will suck the hot air and smoke up the chimney too fast and you'll get no benefit from the fireplace. In such cases, fireplaces have a muffle inside

[11] The heat capacity of air (just 0.7 J/kgK) is less than 2 ten thousandths of that of water (4200 J/kgK) or the walls so it's easy to warm it up as it flows through the house.

the chimney, which must be partly closed in order to restrict the air flow and allow the fireplace to heat up properly.

By the way, Bernoulli's principle is also the reason why, during a hurricane, people board up their windows and doors or leave them wide open. The hurricane's winds may be strong enough to create very low pressure against the surface of the house, and this has been known to suck out windows and furniture, and even lift off roofs, unless there is some air flow to reduce the pressure differential.

Diesel cheating

The expansion of the use of diesel-burning cars has been an absolute environmental and health disaster everywhere. Under the pretext that diesel combustion produces less carbon dioxide (which is doubtful anyway, considering the lengths some manufacturers went to to conceal the real test results), they were offered special incentives, disregarding the fact that they produce much more of nearly all the other dangerous pollutants (NOx, fine soot, etc.) than petrol-driven vehicles. The excuse that they burn less fuel per km does not cut it at all. Diesel-burning cars have been a disaster everywhere.

The sooner they get banned and replaced by electric or at least hybrid vehicles, the better for our health and the environment.

Before we leave the extractor fan, a quick word about electric motors in general. Of course, motors are used in many other places in the kitchen and the home in general, including blenders, cutters, kneaders, the rotisserie in the oven, the compressor in the fridge, the cooling fan in your computer, etc., but this is as good a place as any to discuss their operation briefly.

The electric motor was one of the first, and most useful, applications of electromagnetism, even in the nineteenth century, long before Faraday, Lorentz, Maxwell, and others had managed to understand how electromagnetism works. Essentially any current-carrying wire will create ("induce") a magnetic field around it. This is the opposite of the induction heating we discussed earlier, where the magnetic field also induces a current in a nearby wire. If this wire is now placed close to a magnet (or another magnetic field), they will push either towards or away from one another, depending on the direction of the current and the magnetic field. By wrapping the wire in clever ways around an armature inside a cylindrical magnet and supplying current to the wire, we can make the armature rotate. Nowadays, electric motors are everywhere, not least in the electrically-operated cars that are gradually replacing the polluting petrol or diesel fuelled vehicles.

Digressing just a bit further, I should mention again that generators use the exact opposite principle of the motor: a moving magnetic field induces a current in a wire moving through it. By rotating the wire-carrying armature by any independent means: wind, steam, falling water, etc., inside a magnetic field, we induce a current in the wire which can be used to supply current to wherever it is needed.

Proving Its Mettle: The Metal Pot and Pan

In the kitchen, we use a wide range of different materials, each with a particular set of properties. We use tough steels for pots, pans, knives, and forks, hard ceramics and glasses for cups and plates and for glasses for drinking and for vases, and polymers (plastics) as convenient containers. We even use wooden or bamboo spoons for stirring and as mats to protect surfaces from hot materials because dry wood and bamboo are such excellent thermal insulators and can withstand up to more than 300°C.

So, what are the particular sets of properties that make each of these materials ideal for their function and how do the various alternatives compare? Let's first look at the metals we use. For pots and pans we need quick and uniform heating of food, so conduction of heat and the ability to be heated on the hob or in the oven to a few hundred degrees without damage is important. At the same time they must distribute heat well and must not stain, they must be easily cleanable, they must not scratch, bend, chip, or crack easily, and they must have a long life.

All these attributes can be found in various types of stainless steel, which is mainly composed of iron, a small amount of carbon (about 0.2%), which provides hardness, and also chromium and nickel, both of which make it resistant to rust from oxidation, especially important in the kitchen, where we use salty water.[12] Stainless steel is also shiny and smooth and can be fairly easily formed into pots and pans. Some types of stainless steel ("ferritic") have quite high electrical resistance, so they can also be used in inductively heated hobs.

[12] Salt in water dissociates into Na and Cl, which increases the electrical conductivity and helps to accelerate the galvanic (electricity-assisted) oxidation of most metals.

Iron and carbon

Although hardened iron has been known and used for thousands of years (probably produced by accidental incorporation of some carbon from the wood fire during forging), it was the modern discovery of the hardening effect of carbon in the eighteenth century that ushered in the industrial age. The ability to produce cast iron and steel reliably, together with the earlier invention of the steam engine, are probably the two greatest technological milestones of the first industrial revolution, the basic principles of which are still used to this day. Nowadays, steam engines don't pull trains, but they produce electricity driving generators.

The main difference between cast iron and steel is the amount of carbon. Cast iron contains as much as 3%. This reduces the melting temperature and makes it pourable, which helps to produce detailed products by moulding. It is hard and strong, but very brittle, like traditional ceramics. Everyday examples are manhole covers in the street. There are still eighteenth century bridges in Britain and the US made with cast iron.

Ordinary structural (mild) steel contains much less carbon (about 0.3%), but it also contains other elements, each giving a specific property. There are hundreds of different types of steels.

Stainless steel doesn't have many disadvantages, but it is heavier than anodised aluminium and sometimes suffers from pitting oxidation when exposed to a direct coal fire for too long. This means that stainless steel should not be used for barbeques and similar cooking operations, or even directly on dirty hot hobs, if we wish to avoid this type of corrosion. For electric hobs, only thick-bottomed stainless steel pots should be used, to ensure good distribution of heat. As we discussed before, because its thermal conductivity is moderate, if its base does not make perfect contact with the hob, heat distribution in the pot may not be uniform, and this can affect cooking and warp the pot. For the same reason some drier foods (e.g., eggs or low-fat meat) can overheat at the bottom and easily adhere to it if not stirred frequently.

To speed up heating and ensure excellent heat distribution, many frying pans are made of anodised aluminium alloys, many with a ceramic inner coating. This is because aluminium conducts heat much faster than steel and it's also much lighter. However, even the hardest aluminium alloys are much softer than steel, so they scratch, and deform very easily. They also bend easily, as their elasticity modulus is lower. Moreover, all aluminium alloys soften at less than 550 °C, less than half the melting temperature of steel. Aluminium pans should not be used without an internal coating, as aluminium degrades quickly and even reacts with some acidic foods containing vinegar, etc., which leads to food sticking very easily to it. Internal coatings may be of many types, but the best-known is special high-temperature PTFE ("Teflon"), frequently containing a very fine, hard ceramic powder to increase resistance to wear. Such a composite coating offers very good scratch resistance and a non-stick surface, and it also increases the rigidity (bend resistance) of the aluminium base. In addition, you can safely use steel spatulas and stirring spoons without worrying about scratching, as happens with PTFE-only coatings.

Aluminium is also highly susceptible to oxidation, but it has a trick up its sleeve. The oxides formed on its surface are not swollen and flaky like iron rust but adhere very well on the metal surface, forming a protective barrier. Once formed, no more oxidation of the metal can take place since the oxide is completely impervious and stable.[13] If it gets scratched a new protective oxide layer forms immediately. This is not possible with ordinary steels, which form rust which flakes off. This good property has made aluminium alloys the material of choice for many atmospheric applications such as window frames.

Copper and harder yellow brass (a copper–zinc alloy) or even bronze (a copper–tin alloy) are also occasionally used in the kitchen, but they have the same problems as aluminium, as well as being much heavier. They have excellent heat distribution properties, but are much softer than steel and heavy. Moreover, food sticks easily to them. They are also susceptible to corrosion by acidic foods and discolour easily.

Cast iron, also called "black iron," is iron with a higher percentage of carbon than in steels, making it easily mouldable for things like pans and manhole covers. It was the original, rather primitive steel used for cooking and is sometimes preferred for skillets and frying pans as it retains heat well and is ideal for searing meat, if that's what you like. That's about the only advantage it offers. It's very porous and needs "seasoning," meaning treatment with some cooking oil which polymerises and cross-links, thereby blocking off the pores and coating the metal. It has low to moderate thermal conductivity and is susceptible to corrosion. It is also very brittle, which necessitates excessive thickness and thus increases its weight. When cleaning cast iron, you have to be careful not to remove the "seasoning," which also occurs when you boil water or a stew or use any acidic agent. If the metal is exposed, you'll get a metallic taste in your stew, especially if you use lemon or vinegar, hardly tasty. I just can't understand why some people prefer it to stainless steel or anodised aluminium, unless it is perhaps the "retro" feeling.

[13] It is actually a fine ceramic: Al_2O_3. Thickening this protective oxide layer is called anodising.

The Beauty of Ceramics and Glass in the Kitchen

What about ceramics and glass? Glazed pottery ceramics[14] made from fired clay have been used in the kitchen since pre-historic times, long before metal utensils were developed. They have very good heat resistance (simple pottery ceramics are fired at up to about 900 °C) and I think food tastes better cooked in them. However, because of their brittleness[15] they are often made quite thick and can be moderately heavy. They take some time to heat up and do not distribute heat very well (low thermal conductivity), so they cannot be used on electrical hobs but only over a fire or for baking in the oven. When they do heat up, however, they can hold the heat for longer and so can save some energy. Unglazed pottery needs "seasoning," just like cast iron, to close the pores.

The porosity of unglazed clay ceramics gives them an unusual property utilised for centuries to make a primitive fridge. In hot weather, water absorbed in the pores from inside slowly diffuses out and evaporates. The latent heat used to enable that evaporation is removed from the clay, cooling it down. It also works well with thick canvas and leather. Try it and you'll be surprised how effective it is, especially in hot weather.

Porcelain (also called "china") is another type of ceramic and can be found everywhere in the kitchen, since plates, bowls, cups, and mugs are mainly made of it. There are many different types (one called "stone wear" is particularly strong) and all are made of mixtures of various oxides fired at very high temperatures (over 1250 °C), which makes them very hard. Because it is slightly porous, all porcelain is also glazed during firing to seal the pores, otherwise it would stain as soon as any food came into contact with it.

---

[14] Simple pottery is porous, so it is coated with glaze to close the pores and reduce food absorption by the fired clay. An early Chinese invention, as is porcelain.

[15] They are also brittle, i.e., they have very low toughness and sensitivity to the tiniest flaws, making them unreliable at high stresses.

Hard but brittle

Ceramics are compounds of various metals with oxygen, nitrogen, carbon, boron, etc. Ordinary pottery contains mainly oxides of silicon, aluminium, and magnesium. Porcelain also contains kaolin and other "fluxes" to give high density. Glasses, ice and diamond are special types of ceramics.

All ceramics bend with great difficulty because their atoms are bonded by very strong ionic and covalent bonds, which means that they have very high elastic ("Young's") modulus and dislocations can hardly move to relieve stresses, making them very hard. This also makes them brittle. Compare this with metals which can deform or bend "plastically" without breaking, due to the easier movement of atomic dislocations, which can relieve applied stresses.

Pure or "fine" (or "advanced") ceramics have very special properties and are used in microelectronics, in engines, in batteries, as catalysts, as pigments, in space technologies, as tools and in very high temperature applications.

The main problem with all ceramics and glasses is their low toughness, which makes them very brittle. They do not bend but break if dropped. This is their main disadvantage compared to metals, which deform when dropped and rarely break, as I'll explain a little later too . But have you tried eating out of a metal plate? It tastes metallic. That's because metals react—ever so slightly—with many foods, which slightly alters the taste of the food. Try it. Eat a bit of soup with a steel spoon and then eat the same soup using a ceramic spoon (like those you get in Chinese restaurants). There is a definite difference in taste, at least I think so. The reason is probably the nickel contained in stainless steel, which is a powerful catalyst, but iron is too. It is also the main reason that cheap steel knives tend to brown fruit quickly and can affect the taste of some fruit and vegetables when cut. This happens because of accelerated oxidation (reaction with oxygen) in the presence of nickel and iron. As I mentioned before, there are good ceramic knives (made of a fine "zirconia-toughened alumina") which reduces this effect.

Plastic Fantastic ... But Not All Plastics Are Born Equal

Look around you. I bet you can see dozens of plastic (properly called polymers) objects, especially in the kitchen. The vast majority of food containers are made of different kinds of plastics. And they are all made by cleverly guiding and bonding millions and billions of molecules together in a huge number of long, millipede-like macromolecular chains, all coiled and wrapped up together.

But polymers are everywhere in nature too. Do you know which is the largest organ in the human (or any animal) body? It isn't the liver or the intestine or the brain. It's the skin. It is a huge, tough system of different polymer chains mixed and working together: collagen, thousands of different proteins, hyaluronic acid, even the DNA in the cell's nuclei. Our skin acts just like the film you use for covering dishes or wrapping your sandwich, protectively covering every inch of your soft, easily damaged muscles and organs. And by the way, the protective "skin" (i.e., the paint) enveloping any kitchen appliance (or a car or an aeroplane or anything painted for that matter) is a polymeric material exactly analogous to the skin of an animal.

The main characteristic of a polymeric material is that it is flexible, certainly more flexible than nearly all metals and all ceramics and glasses. That's because the coiled and wrapped macromolecular chains can easily bend and slide over one another. The units (modules) of a macromolecule—monomers, small, non-repeating molecules—are joined together by hydrogen bonds which depend on electrostatic attraction between the atoms of adjacent monomers. On the other hand, the bonds *between* the macromolecules are very weak (called "van der Waals" forces from their discoverer). This allows an easy sliding movement between the macromolecules. That's why we can flex and stretch our skin and muscles and that's why most plastics can bend, stretch, or twist to a large extent without significant damage.

As we saw before, cooking creates many polymers, be it reduced sauces or starch-based macromolecules. The way we do that is by imitating nature. As we discussed, nature utilises specialist molecules with some trace metal elements as enzymes to catalyse the resulting polymeric proteins and we try to repeat this in the pot. We add lemon or alcohol or vinegar or egg, which are also catalytic and help bring oils and water and other molecules together to create small chains of mixed polymers: the sauces and emulsions that make all the difference in a dish.

Traditional plastics (artificial polymers) are made from processed petroleum oil and are therefore fossil-based materials and not sustainable. For this reason, there is a strong incentive to find sustainable plastics made from plants, making sure that they can be recycled too. Let's first look at the most important types of plastics used in the kitchen and then we'll talk about biomass-derived plastics.

There are two main types of polymeric materials made from fossil-based oil or gas. The most ubiquitous are the thermoplastics, all of which melt first when heated and can therefore be recycled and also processed fairly easily. Good examples are various types of polyethylene (PE), which are used to make cheap plastic bags and cling films for wrapping (low-density

PE) or stronger plastic bags and cheap liquid containers (high-density PE). Plastic gears, wheels, and women's stockings are made of Nylon (the first soft plastic invented). Water or drinks bottles are made of clear polyethylene terephthalate (PET), while polypropylene (PP) is used for many modern food containers and plastic bottle caps, replacing polyvinyl chloride (PVC) which was the previous material of choice for containers and has now all but disappeared. Polycarbonate (PC) is used for strong containers and for protection and impact shields. Acrylonitrile butadiene styrene (ABS) is used for crockery, while polystyrene (PS) is used for cheaper packaging and single-use utensils. Poly-methyl-methacrylate (PMMA, a type of acrylic, sometimes known as plexiglas or perspex) is used extensively for hard containers, as it looks and feels a lot like glass. There is also thermoplastic polyurethane (TPU) which behaves like a rubber and is used for seals and some utensils. Most of these—and more—plastics are used to produce thousands of different products by melting and extrusion under various conditions for specific situations. Many are mixed together (blended) during production to create blended materials for unusual applications. It is also possible to join them together to create composites, such as the multi-layered film membranes with very good oxygen barrier properties that are used for packaging sensitive foods. All these thermoplastic materials can and should be recycled, and nearly all can be washed and re-used again and again for various jobs in the kitchen.

The second main category of polymers is the thermosets, many of which are also used in the kitchen. Examples are certain glues (epoxies) for seals and flanges, Bakelite for insulation and wall sockets (the very first hard polymer), hard polyurethane (PU) for foamed insulation, polytetrafluoroethylene (PTFE, frequently known as Teflon), used for crockery linings, and elastic rubbers (EPDM, silicone rubber) for seals and utensils. Products made from these plastics rely on moulding and compression of a catalyst-activated resin. Unfortunately, none of them can be recycled as they do not melt when heated. Instead, at high enough temperatures, they decompose and burn, producing dangerous fumes and carbon.

Plastics are extremely versatile and stable materials and generally safe, but can only be used safely at relatively low temperatures compared to metals and ceramics (at most 120 °C for thermoplastics and up to 200 °C for thermosets), so they cannot be used to cook food. Probably everyone has had the unfortunate experience of a plastic dish melting or catching fire when exposed by mistake to a hot surface or a flame. In such cases, the fumes released contain very dangerous gases with a large proportion of aromatic hydrocarbons, a group of carcinogens.

Certain plastics are resistant to relatively high temperatures and are used accordingly in the kitchen. These include PTFE, used on its own or mixed with some ceramic powder (a composite) for lining frying pans and pots, and silicone rubber, which is used for oven moulds and stirring spoons and scoops.

Industry assures us that plastics are safe for storing liquids and various foods, and this is largely true, as leaching from most modern food-grade plastics is very rare. Specifically, modern high purity PP and PET are designed to be safe for storing cooking oils and drinking water. However, there have been various reports that plastics we have all grown up with such as soft PVC (the ubiquitous well-known soft containers produced for decades and sold until recently) have been found to leak out a dangerous substance called phthalate. This was used to soften the normally hard PVC. Let us hope that such serious problems have made producers and authorities more careful and thorough in testing the safety of food-related products.

Nearly all fossil-derived plastics are extremely resistant to chemical attack from acids or bases. Most are also highly resistant to ordinary organic solvents such as ethanol (alcohol), but a few (e.g., PMMA) can be attacked and dissolved by stronger solvents such as acetone and chloroform. Many are also resistant (or can be made resistant) to ultraviolet radiation from the Sun, which often hardens plastics by initiating cross-linking between macromolecules, resulting in loss of transparency and cracking due to brittleness. Some, such as PTFE and food-grade PP, seem to be oblivious to everything, including the strongest organic solvents we can throw at them. These plastics seem to have lifetimes measured in centuries or even millennia before they are broken down in the environment. Most of the others still need many years or decades to break down.

This is then the main problem with plastics. Being cheap, light, versatile, incredibly stable, and resistant to chemicals is both a blessing and a curse. It has meant that the use of plastics has spread everywhere—especially as packaging and as single-use utensils—over the past decades, while at the same time plastics discarded in the environment have dispersed all over the world. Many of them break down by natural processes into tiny microplastics which cannot be broken down further and therefore accumulate in the environment. For this reason there is a ban on the production and use of single-use plastics, spearheaded by the European Union, and a parallel effort to find replacements that can be recycled or degraded within a reasonable time in the environment.

Very few useful strong biomass-based plastics have been developed till now. The most promising and already widely produced is polylactic acid (PLA), which is made from corn or potato starch (a natural polymer) and recently widely used for biodegradable plastic bags and single use cups and straws, as well as some niche applications. It is a very strong and very durable thermoplastic and can easily be recycled, but its relatively low use temperature (below 55 °C) is its main drawback. This means that it cannot easily be used in the kitchen, except as non-heated or fridge containers. Recent developments have meant that wrapping films can also be made of PLA or blends.

Cellulose acetate and similar bioplastics have been around for many years and are derived from wood cellulose as well, but they are weak and can only be used for low grade packaging of foods. Most bioplastics degrade in the environment under ordinary conditions.

Cracking and Scratching

Mechanical damage of materials can occur in many different ways. Stretching (called tension in physics), pressing or scratching (compression), twisting (shear), or bending (flexure) are all loading modes and can all result in damage to the material being loaded, but also to the material applying the loading. In the kitchen, such damage occurs in many operations between materials so let's consider some of the more usual—and unusual—situations.

First of all we must remember that it is not the total force we apply (with our hand, for example) that is important when we discuss mechanical damage, but the "stress" that the materials feel. This is simply the force (we call it the "load") divided by the area upon which it acts. The smaller the area, the higher the stress felt by the material. This is very nicely demonstrated by a person wearing sharp heels as compared with the same person wearing trainers. In the former case, the stiletto heels will exert a large stress and easily dig into the ground, whilst the trainers will only leave a slight mark.

Now let's see what happens when we are pressing one material against another. When loading begins, both materials behave "elastically," which means that their atomic bonds simply stretch and deform without breaking, and the materials will go back to their original shape if the load is removed. After a certain deformation though, some bonds start to break. If they reform at a different position, we have "plastic deformation" of the material, which means that, upon unloading, the materials will show a permanent deformation, perhaps a groove or a scratch or a curve. The amount of plastic stress that each material allows is called its "hardness". Generally, plastics have low hardness (their coiled macromolecules can easily uncoil and stretch), while ceramics (with very strong bonds) have very high hardness. Metal atoms are joined by moderately strong bonds, so their hardness lies in-between. Hence, most metals and polymers deform plastically and are said to be "ductile," whereas ceramics, glasses, and cast irons cannot easily deform plastically and are said to be brittle.

At much higher deformations, all materials will eventually break as the atomic bonds cannot deform or re-form any further. A final fracture can take place suddenly and all at once, as in ceramics, glasses, and cast irons, or after the formation of many small microcracks, as in many plastics and metals. The stress at which final fracture starts is called the material's fracture strength.[16] Since each bond absorbs a certain amount of energy before it breaks, the total energy required to break all the bonds till final fracture is called the "toughness" of the material. Generally, plastics have low toughness while most metals have high toughness. Ceramics, glasses, and cast irons have low toughness; they break soon after maximum elastic deformation, because they lack the capability of sliding atomic dislocations to relieve stresses, as we discussed earlier. Some high tech ceramic composites have moderate toughness.

Well, that's the theory. Let's see what happens in practice in the kitchen, for example, when we are cutting a piece of meat with a sharp knife.[17] Because the area upon which we apply the force is very small (the sharp edge of the knife), both materials will feel the same high stress. The material that will be cut first is the one with the lowest hardness and strength, in this case the meat, after much elastic and plastic deformation of the atoms in the cellular polymer proteins. But what happens when we try to cut a piece of meat on a ceramic or glass plate? As soon as the meat is cut, the knife finds itself pressing against the much harder ceramic. In this case, the metal's atomic bonds will give in first and the metal's edge will be permanently deformed, although hopefully not fractured. The metal knife is now deformed quite a bit and will have to be re-sharpened,[18] while the ceramic plate will not show a scratch.

[16] Fracture strength is generally different depending on loading mode. In metals it changes little, but in ceramics it is much higher under compression than under tension or flexure, so buildings are built of ceramic bricks and concrete with steel bars as reinforcements—the best of both worlds.

[17] The sharpest knife or surgical blade is still many atoms thick at its edge.

[18] Sharpening is done by rubbing against a harder metal or a ceramic.

Beautiful, hard, and cold

Diamond, the hardest known natural material, is a crystal allotrope (different atomic arrangement) of carbon, the other being graphite with multilayers of graphene. Diamond's extreme hardness is due to the dense arrangement of its carbon atoms, which restricts movement of atomic dislocations. It appears transparent as its electrons have no convenient quantum energy levels to absorb visible radiation, just like glass and water. Synthetic diamond is sharper than natural diamond as it is made under controlled conditions at up to 60,000 bar pressure and 1100 °C. It is also possible to grow amorphous ("glassy") diamond in layers or coatings. Coloured diamonds are due to the inclusion of small amounts of metal ions as oxides, but yellow is due to atmospheric nitrogen.

Diamond is considered to be a ceramic, and like all ceramics, it has low toughness so it is easy to cleave and chip. If encased in a hard metal it can be used for drilling into rock, cutting, and grinding. In addition, it has the highest thermal conductivity of any material (a test of authenticity is to touch it with your tongue. If it feels colder, it is authentic), so it can be used as a thermal sink for microelectronics. It is also an excellent electrical insulator.

What about a knife cutting herbs against a hard wood or plastic cutting board? The herbs will give way first, and the hard wood or plastic will then get scratched or grooved, but it won't break. Here we have an intermediate situation where the hard wood or plastic will cause the metal edge to become slightly blunt, and after many cuts it will need sharpening again. The point is that, no matter how hard and sharp a metal knife is, the stress felt by its cutting edge is high enough to mean that there will eventually be sufficient atomic-level deformation for the knife to become blunt.

An advanced ceramic knife on the other hand (usually made of pure zirconium or aluminium oxide) will survive much longer, as long as it is not knocked around. Its hardness is so high that, even when used against a ceramic plate, its cutting edge will remain sharp for a long time, although not for ever. It will easily scratch even the hardest metal kitchen knife, let alone a plastic board. Only a diamond-edged knife (they do exist and are used in physics) will remain sharp nearly for ever, as long as it is not knocked, which can crack it. The bottom line is that scratching and marking has all to do with the relative hardness of the materials. All materials, except diamond, can be scratched by another material, although breaking a material requires much greater force than simply scratching.

Cold Makes Things Snap

Have you ever noticed how easily some plastics crack and tear when they have been in the freezer, but never when they have been in the warm? When warm or at least at room temperature, the macromolecules are soft and elastic and most can bend easily without breaking. That's because the atoms making up the monomers have just enough energy to allow partial uncoiling and stretching of their bonds without breaking. But as soon as the temperature drops (as in a freezer, where the temperature is generally about − 18 °C), the atoms vibrate much less, so the bonds cannot stretch as much and they break easily, even after a small deformation. The temperature at which this ductile-to-brittle transition takes place is different for each polymer. Some metals also suffer from this problem. In the past, whole ships have sometimes broken in half during winter storms as the simple steels they were made of became brittle.

This effect is much more pronounced for many elastic, rubbery materials but also for PMMA and PP. Under normal conditions, a rubber band can stretch to many times its length. But if you put one in the freezer and stretch it, it'll break after a much smaller extension. That's why you find them broken if you use them for holding packets closed in the freezer. By the way, ozone produced by the sun and fluorescent lights also makes elastics and rubber brittle.

Some plastic containers are not made for the freezer and crack at very low temperatures. PMMA, PP, and PVC will crack easily in the freezer, whereas LDPE and HDPE remain elastic in the freezer and can even be used at much lower temperatures. ABS and PLA are at an intermediate level.

Cold breaks things

Cooling metals (and plastics) to very low temperatures changes the way their atoms can move. Copper and aluminium have a face-centered cubic atomic lattice (atomic arrangement), so there are many ways that their atoms can slide, called slip planes, to allow them to deform without breaking. These are called "ductile" materials and can be used in studies all the way down to (close to) absolute zero (− 273.15 °C).

If the number of slip planes is limited, as in body-centered cubic metals and in metals with distorted structures (carbon steels, brass, etc.), deformation becomes difficult, so these materials become brittle at low temperatures. This ductile-to-brittle transition has been the reason for many spectacular failures in the past, as in the breaking up of whole ships, bridges and pressure vessels in very cold weather.

To defrost anything plastic from the freezer, always allow it to sit for some time on the table before attempting to remove the lid, to avoid cracking it and the container.

From the metals, aluminium and copper are fine for the freezer, but not simple steels. Most stainless steels are fine too, but the pot and pan handles are not made for the freezer and will crack easily. Ceramics, glasses, and cast irons are brittle anyway so they are not affected by being placed in the freezer.

Kettles Like Spaceships Taking Off

How would you tell when the water is just about to boil in the kettle? By listening to it. When the kettle is switched on, the element at the bottom (or the internal resistive element in older kettles) heats up rapidly. As we discussed before, the heat very soon causes some of the water molecules to get enough energy to start evaporating internally, which creates small bubbles at random positions on the hot surface. Because they are surrounded by water, these bubbles grow a little and burst continuously before they reach the surface ("implode"), and each makes a small but high-pitched noise. The bubbles grow rapidly and, once large enough, their buoyancy is sufficient for them to come unstuck from the bottom and rise to the surface by buyancy, whereupon they release the steam they contain.

As more and more heat is pumped into the kettle, more and more bubbles form and the noise level increases. At the same time, the water at the bottom gets hot and, since it is now lighter, it rises from the bottom to the surface where it cools down and sinks again causing the familiar boiling motion which ensures that the hot water is continuously mixing with the colder water. As boiling temperature is approached (depending on the altitude), the water mixes better and the heat becomes better distributed throughout. Fewer, but larger bubbles form at the bottom and the noise becomes deeper and less intense. Finally, the water begins to boil and mix violently, and the few large bubbles that form no longer implode inside the water, but quickly rise to the surface and release their steam, at which point the noise abates. That's when you know, without looking, that the water has boiled.

Some kettles are much noisier than others, depending on how good their soundproofing is, the material they are made of, and their overall design. Plastic kettles are generally quieter because plastic is soft and tends to deform (low elasticity modulus), thus damping the sound waves. Metal kettles on the other hand are much louder as the metal has high elastic modulus and cannot deform enough to absorb the sound waves. The quietest kettles are ones that are made with double walled plastics which almost completely absorb the sound waves. Old-style kettles used to have a whistle at the outlet which certainly let you know when boiling has commenced!

Most modern kettles have an automatic shut-off mechanism in the kettle when it reaches boiling temperature. This is actually an ingenious mechanical thermostat (we saw a similar one earlier in the case of the toaster). This is set to open the circuit when the water temperature reaches 100 °C, cutting off the current and stopping the boiling. It works by cleverly combining thermodynamics (in this case, the thermal expansion of two different metal strips stuck together) with the spring-like elasticity of hard metals. If you ever take apart a kettle, you'll be amazed how cleverly the thermostat has been designed in order to take advantage of these laws of physics. Apart from toasters and kettles, such "bi-metallic" or bent single strip thermostats are used in clothes irons, stoves, ovens, pressure cookers, fridges, boilers, hair driers, hair straighteners, and many other devices. It is one of the most useful and ubiquitous little devices ever made!

An Igloo in Our Home

A fridge (or two) is one of the few appliances that are absolutely necessary in any kitchen for keeping food fresh. Let's first see how a fridge works. The basic idea is to remove heat from inside.

The first thing to note is that, by trying to expel heat from a cool place (the fridge) to a warmer place (the kitchen), we must violate the second law of thermodynamics, which says that heat can only flow from a hot to a cold place. Because this law is paramount, the only way to succeed is by putting a lot of energy into the system and using it to "push against" the second law. We'll see that this is done by a compressor and uses up quite a bit of electricity.

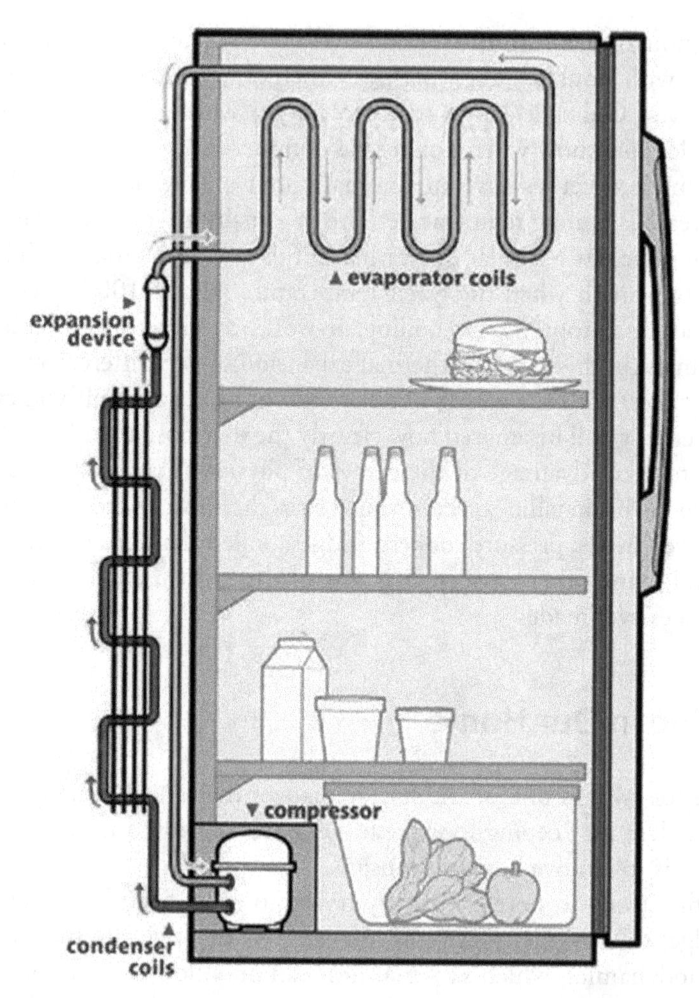

Schematic of fridge operating principle

The fridge simply exploits the latent heat of vaporisation as a liquid transforms into a gas. The fridge has a network of pipes embedded in its walls and the insulation all around it, but especially in the freezer compartment. This network of pipes is called the "evaporator". A second network of narrow pipes sits behind and outside the fridge chamber and is called the "condenser" (see figure). Inside these pipes flows a "cooling liquid" that transforms to a gas at a low temperature and has a large latent heat of evaporation—we also saw this property when discussing water. This means that, in order for it to transform from gas to liquid, it must release a lot of heat. Conversely, it must absorb a lot of heat to convert from liquid to gas, and this exactly what we need in a fridge.

Freon kills ozone

For normal fridges, the coolant gas used is coded R134a. It is the replacement for Freon 12, responsible for almost destroying the ozone layer (dissociating the O_3 molecule) in the upper atmosphere, which protects us from the Sun's UV-C rays. Freon was also widely used for producing expanded polystyrene ("styrofoam"). In 1996 Freon 12 was banned worldwide and the ozone hole has since started filling up again, in a rare success story for global collaboration. Unfortunately, there are signs that someone somewhere has restarted producing Freon, as the hole has recently been opening up again.

The boiling point of R134a (when evaporation occurs) is about − 26 °C. That's why the best (normal) freezers can only go down to about − 25 °C. Freon could go down to about − 30 °C.

So what happens in the fridge is this: the gas is first compressed by an electric compressor pump and, because of the ideal gas law we saw earlier, its temperature increases (the volume V is constant so T will go up as P goes up). It essentially picks up energy from the electricity driving the compressor. It is then fed to the condenser, where the heat is dissipated to the kitchen environment and the gas cools down to become a liquid at high pressure. This is then fed into the evaporator in the fridge, where the liquid is allowed to expand (in an expansion device) and transforms into a gas again, absorbing heat from the fridge because of its latent heat of vaporisation. This gas is then returned to the compressor to start the cycle again. That's it. A continuous evaporation–compression–condensation cycle that removes heat from inside the fridge.

Liquid-to-gas transformation occurs in other systems too. We already saw it in gas bottles. When we use a spray can, we are expelling liquid at high pressure. This transforms to a gas when released and feels very cold. In fact, if you want to cool down something very quickly, just spray it.[19] As I mentioned before, in ancient times and still today in many parts of the world, a simple cooling chamber was made by porous ceramics or other materials (cloth, leather) that allowed water to evaporate in the sun, cooling the chamber by latent heat absorption. I've seen little camping fridges doing the same. Modern "solar fridges" use the same principle.

Interestingly, bar fridges (of the kind you find in hotels, closed in a cupboard, but many modern ones too) have the condenser inside the fridge wall. This means that the heat of the compressed liquid is dissipated inside, so the fridge air never gets very cold. And the compressor has to work

[19] Many years ago, burglars broke into my home by doing exactly this: they sprayed the brass lock mechanism to make it brittle and then broke it easily, due to the ductile-to-brittle transition we saw earlier. Standard burglar practice, apparently.

almost continuously. A bit of a waste of energy actually.[20] The small camping fridges that use the Peltier effect to cool a small chamber by thermoelectricity are also a bit of a waste of energy. Their efficiency is still very low.

Low temperature slows down all reactions but also increases relative humidity, keeping some vegetables fresh for longer, although other foods require drier air for correct preservation. Air can only hold so much water at a given temperature, so when fresh air enters the fridge, it cools down, but there is now more water than before so the relative humidity of the air in fridge increases. Most modern fridges have systems to try to control and even expel humidity, not always successfully.

On the other hand, we don't want to reduce the humidity too much in case sensitive vegetables such as lettuce, peppers, and cucumbers dry out more than optimally. There is a good reason why vegetable crisper drawers are usually at the bottom of the fridge while cheese compartments are at the top. When we open the door of the fridge, humid air enters the fridge and since it is heavier (higher density), it sinks to the bottom, where the crisper drawer is. Conversely, drier air stays at the top of the fridge, where the cheese drawer is. We certainly don't want too much humidity there, to avoid mould.

Storing leftovers in the freezer (down to − 18 °C or lower) is only a good idea if the food does not contain semi-cooked vegetables with a cellular structure. As we discussed before, in that case the water in the cells will freeze and destroy the cell walls, causing the structure to collapse. Lettuces, cucumbers, peppers, and tomatoes will all become soggy and uneatable. Cabbages, cauliflower, beans, sprouts, peas, and similar vegetables are fine.

Shutting the Door

So, we cooled the inside of the fridge, but the second law of thermodynamics dictates that heat should always flow from hot to cold. So we need to make sure that heat from the warm kitchen does not enter the fridge. We do this by insulating the fridge as well as we can, all around, and keeping the door closed as much as possible.

Thermal insulation materials have progressed tremendously over the years. Fridge insulation is nowadays made of a hard thermosetting porous polymer called polyurethane (PU), which is sprayed on and penetrates every nook and cranny between the metallic exterior and the inner chamber. Such expanded

[20] But nothing as bad as some commercial fridges in summer at the back of old shops, where the heat from the condensers warms up the air, eventually finding its way back inside the fridge. The poor compressors work continuously to no avail.

PU has extremely low thermal conductivity[21] and is ideal here since, in such closed, filled spaces, the only way for heat to penetrate is by conduction. The large number of pores means that there are only very few, thin paths along which heat can penetrate inwards.

The outside of the fridge should always be shiny white to minimize emissivity and hence energy absorption by radiation. Recently, I saw some grey stainless steel fridges, but their shine saves them as it probably ensures low emissivity.

Vertical fridges tend to lose some cold air when you open the doors, since cold air sinks (it has a slightly higher density as the atoms are vibrating less and are thus closer together). This is immediately replaced by warm air coming in from above. However, horizontal floor freezers have no such problem when you open the top door. The cold air inside remains where it is and restricts any warm air from entering. You don't get any smells escaping either.

The rubber seals on vertical fridges are quite cleverly made to give a good deformable, compression seal. But have you noticed that the doors tend to be slightly difficult to open and become a little more difficult with time? That's because a small amount of cold air leaves the fridge and an equal amount of warm air replaces it when you open the door. In addition, the door seal deforms as you push against it expelling some air and reducing the internal pressure. With time, the warm air cools down inside and shrinks, reducing the internal air pressure even further—in other words we create a slight vacuum inside. This pressure difference makes it a little difficult to open the doors, since you now have to overcome some of the atmospheric pressure around you. It's a good thing the seal is flexible otherwise it'd be a battle to open the door.

Defrosting the fridge and especially the freezer is another interesting thing worth discussing. Ice forms on the inside surfaces firstly by condensation of humidity (when you open the door) and eventual freezing of that condensed water. With time, thick ice deposits often form in the freezer and some even in the fridge. Solid ice is not a good insulator (it has a thermal conductivity of about 2 W/mK), but the ice that forms on fridge and freezer surfaces does so gradually over a long time and is of a very porous type. This means that it is an excellent insulator, keeping the inside air from being cooled by the circulating coolant gas in the walls, and eventually increasing the chamber temperature and wasting electricity. And perhaps wasting food. Most modern

21 Around 0.04 W/mK, a little higher than expanded polystyrene. The conductivity of still air is about 0.001 W/mK and that of the vacuum is zero. For comparison, the value for a building brick is about 0.6 W/mK and that of steel about 50 W/mK.

fridges have an automatic defrosting system,[22] but most freezers do not, and it's always a good idea to remove the ice regularly. Horizontal fridges trump vertical ones here too, since they do not allow warm air in, so condensation is minimal.

Since we are talking about warm air in the fridge, a word about filling the fridge or freezer with food. While it's a good idea to store cooked food in the fridge, don't rush, let it cool down sufficiently before putting it in. Putting warm food in the fridge not only creates condensation inside the container (as we discussed previously, risking the development of mould), but also warms up everything else inside the fridge, wasting electricity and reducing the preservation time of all the other foods in it. If you like soft ice cream, help yourself about an hour after putting warm meat in the freezer. This is especially true for nearly empty fridges. In full fridges, the added heat is distributed to a larger mass so this problem is not so obvious.

Finally, a comment on and a solution for fridge odour. The inside surface of all fridges is made of plastic, and all plastics are slightly porous. If there is a smelly food in the fridge, the vapour will gradually diffuse into the pores of the wall and then the insulation, from which it will not escape easily, because of its own surface tension. Small molecules can also adsorb[23] onto the surface of the plastic walls. All fridges suffer from this problem and the door should never be closed when they are switched off and empty, to allow the smell to gradually dissipate. A way to remove smells is to beat the fridge at its own game. Place a small saucer of finely ground coffee in the fridge (very fine grained Greek or Turkish coffee works well, and I think espresso too) and refresh it every week or so. The fine grains have nanosized pores and a huge specific surface area.[24] They will preferentially absorb all small vapour molecules in the air, successfully removing smells and even gradually encouraging desorption of most smells from the walls.

A Non-sticky Condition

When we talked about frying I didn't mention why food sticks to the pan and how non-stick pans work.

Generally, satisfactorily pot-cooked food has polymerised fully and the oil and water (and nutrients) have emulsified successfully. In this form, they have

---

[22] Any ice forming on the back wall inside the fridge is thawed regularly and fed onto a tray sitting on top of the hot compressor behind the fridge, where it evaporates away.

[23] Gases "adsorb" onto a surface but "absorb" into a volume.

[24] Once I measured the SSA of fine Greek coffee to be about 100 m^2/g.

very low affinity to metals or glazed ceramics or glass, so they should not adhere at all to the pot. However, if any of the food has been overheated, e.g., by inadequate stirring or allowing the liquid to reduce too much, it will have no free water and will have transformed to acrylamide and other compounds that adhere strongly to metals and glass. Semi-glazed ceramics and cast iron have particular problems as they are very porous and food will diffuse through the pores, creating particular adherence problems and sources of contamination that can alter the taste of the food. Even stainless steel is not immune to sticky, jellified foods.

Frying pans and oven dishes have additional problems because polymerised oil or fat vapours may condense on the outside as well, and if they are not removed quickly, they will cross-link over time, creating hard polymers that adhere extremely well on the outside. Only strong, caustic materials and scrubbing at high temperatures can removed such deposits.

Finally, a word about non-stick pots and pans. These are generally coated with PTFE, a thermosetting polymer which has high temperature resistance (up to 280–300 °C) and a very low friction coefficient. In fact, it is one of the most difficult materials to stick anything on. For this reason, you can easily fry or bake anything without the use of oil, which is used to reduce adherence by reducing surface tension to other materials. As I mentioned above, PTFE is a relatively soft material and scratches easily so in order to reduce its wear, modern frying pan coatings are made by combining PTFE and a hard ceramic powder (usually aluminium oxide, Al_2O_3).[25] But, if by any chance, a PTFE or PTFE + ceramic pan is left on the hob and overheats to above 300 °C, and if it is visibly damaged (you'll probably see some fumes and smell the mess), then it is probably dangerous and a good idea to replace it.

When Hot Meets Cold—Shocking Heat

We've already talked about cold. But what happens when cold meets hot? For example, when we take out a hot glass dish from the oven and place it on a cold surface, say a cold hob.

It's a terrible idea. Even if the dish is made of stainless steel, it will shrink slightly (thermal shrinkage) immediately over the area of the cold surface while the material around it will remain warm. This will set up large stresses in the material which will probably end up warping it. Over time the

[25] Misleadingly called "ceramic-coated". Older enamel-coated metallic pots were made using a glass + ceramic mixture, but the enamel was also brittle.

warping will become permanent. Aluminium and copper warp worse than other metals from such "thermal shock".[26]

Glass or ceramic or cast iron oven dishes won't be so lucky. Even an oven-proof glass dish ("pyrex") may crack from these thermal stresses. As we discussed before, ceramics, glasses, and cast iron are all very brittle materials. They tend to crack when subjected to large stresses since they cannot bend or warp to accommodate the stresses set up by the differential thermal expansion. It's actually worse than that. Their thermal conductivity is lower too, so they cannot immediately dissipate the thermal stress by spreading the heat around, increasing the local stresses even further. These materials have a low "thermal shock resistance," which is about 180 °C at best for ceramics and a little more for pyrex glass.[27] This means that a glass or ceramic or cast iron casserole dish must only be placed on a dry[28] wooden or cloth surface (which are both full of air and therefore good insulators) when removed from the oven to avoid possible cracking.

Since we are discussing stresses in glass, I'd like to make a comment on the so-called "unbreakable" ("safety") glass tables, doors, walkways, etc. Unfortunately, they are certainly not unbreakable, as many people have found out to their cost.

Such glasses are properly called "tempered" and are made by an interesting process. They are floated as any ordinary glass pane,[29] but then put through a fast cooling process in which the outside surface "skin" (top and bottom) ends up under compressive stresses. This means that any loading (say a hammer knock), that would easily crack an ordinary glass has no apparent effect because the compressive stresses cancel out the tensile stress applied by the hammer. So far so good and the tempered glass survives most knocks. But there is a catch. The surface compressive stresses are always balanced by large tensile stresses in the core of the glass.[30] If by any chance a small crack penetrates the surface into the core (say at a surface flaw growing slowly), the result is an almighty explosion as the internal tensile stresses are suddenly released and the glass shatters into smithereens.[31] Exactly the same can happen if

[26] Their coefficient of thermal expansion is higher than that of stainless steel.

[27] For comparison, most metals used for cooking have a thermal shock resistance of more than 600 °C, although they will warp when subjected to such an extreme temperature shock.

[28] If the wood or cloth glove gets wet, its thermal conductivity is 100 times higher, so you'll burn your hand if you use it.

[29] To make it perfectly flat with equal thickness, the glass is poured and floated on a bed of molten metal tin (Sn) and gravity looks after its flatness.

[30] Remember all forces and stresses must be balanced out. They cannot exist on their own, but always occur in pairs.

[31] Because they are small, these pieces are not as dangerous as shards. Unless you happen to be right there during the explosion, I guess.

such a glass is thermally shocked. So, no, tempered glass tables, etc., are not unbreakable and should still be treated with respect just like any other glass or ceramic. By the way, because this problem is well known, many such glass items are (should be) coated by strong plastic sheets (usually polycarbonate) top and bottom to contain the particles of any explosion. If you really want impact-resistant glass then you can use armour plate glass which is a sandwich of glass with polycarbonate sheets.

The Kitchen was Hi-Tech Before 'Smart' Tech Came Along

There is lots of technology in a modern kitchen. Sensors, dials, control panels, automatic switches, buzzers, touch screens, etc., abound. In reality though, to cook properly, we don't need such technological marvels. We only need to know the temperature of the food and the temperature of a stove or a hob. All the rest are helpful but optional. I'll only look at how we measure the temperature here and discuss a few other devices later.

We have a simple way to determine the amount of heat energy in a material or the environment: at low enough temperatures, as encountered in cooking (up to about 250 °C), we used to use a glass capillary thermometer where we measured the amount of thermal expansion of a liquid (mercury or alcohol) in a thin capillary, an expansion which occurs due to the vibrations of its atoms. Nowadays, we use electronic thermometers, which work very differently: they convert the heat they sense directly to electricity via a number of physical phenomena. The electrical resistance of a platinum sensor increases as the heat increases and this gives us the temperature, after calibration against another known thermometer. A very similar type is the thermistor electronic thermometer, the main difference being that increasing temperature *decreases* the resistance of the thermistor. Yet another kind of thermometer uses a thermocouple (a joint between two dissimilar materials), which exploits the Seebeck effect (the inverse of the Peltier effect) and has been the mainstay of temperature measurement in the science lab and the kitchen ever since its discovery.

Similar electronic thermometers are used for measuring the temperature inside the fridge and freezer. In addition, meat thermometers utilise the same principles, with a sensor at the end of a long sharp pin which can be inserted into the food.

Finally, we have the infrared, non-contact thermometers, which utilise the principle that different wavelengths of light (in the infrared region of the

EM spectrum) correspond to different temperatures. The problem with these thermometers is that they only measure surface temperature, and also that they involve knowing the emissivity of the body in order to compute its temperature correctly. We saw earlier that the emissivity of a shiny body (like a roast covered by oil) is very low, while a dry roast would have a high emissivity, misleadingly appearing hotter. So measurements of wet or shiny surfaces with an IR thermometer cannot be relied upon.

6

Physics is Everywhere You Look and … Hear

"No one undertakes research in physics with the intention of winning a prize. It is the joy of discovering something no one knew before."

Stephen Hawking

G. Vekinis, *Physics in the Kitchen*, Copernicus Books,
https://doi.org/10.1007/978-3-031-34407-7_6

Well, I hope you agree that physics is indeed everywhere in the kitchen, whether we are considering the pot or the food itself or the appliances around us. No matter where we turn, we can't avoid the fact that the basic make-up of the different ingredients, the way they respond to cooking, and the way various devices operate are all driven by physical principles.

In this last part of the book we'll have a look at various aspects of our life in the kitchen that are "tangentially" relevant to our cooking and operating in the kitchen. And we'll start by listening to the kitchen symphony (or cacophony, more correctly, I guess).

Kitchen Symphony

Earlier we saw that many kettles make a racket simply by the creation, development, and bursting of water bubbles. But the kitchen is full of many other noises during cooking, producing a veritable cacophony of sounds. Let's investigate some of the sources.

Pot cooking produces only a limited amount of noise since the bubbles produced are muffled by the surrounding food. Occasionally, a tight-fitting lid may emit high frequency noise due to escaping steam.

The loud noise you hear during frying is from a different reason to boiling. It occurs when superheated water bubbles explode. Deep frying involves firstly heating oil to a temperature much higher than that needed for boiling water, generally above 150 °C. When a piece of potato or chicken or fish or meat is placed in the hot oil, its own water is suddenly overheated and evaporates immediately, forming small bubbles of steam. As it is inside the oil and expands suddenly, the steam in the bubbles has nowhere to go and increases its pressure until it either implodes in the oil or explodes at the surface, releasing its energy to the surroundings, together with a fine oil spray and noise. When the accessible water contained in the food is completely consumed (mainly the surface water), the temperature rises quickly and can soon affect the food itself, resulting in the familiar acrylamide and lower frequency noise. Shallow frying sounds have much higher frequency, as the small bubbles are close to the surface and are able to burst at a lower pressure. Again, as the water is consumed, the sound subsides a little.

Sounds from oven cooking are very limited since the door tends to dampen the sound, unless the fan is being used. As the air extracted is hot and has lower density, the sound is at a lower frequency.

A very loud sound is sometimes made by resonating cavities, as in my hand-held mixer at specific speeds. What happens is that, under certain conditions, the vibration of the plastic of the mixer by the rotating gears happens to reach its "natural vibration frequency" and, being an almost closed cavity, it sets up standing waves, just like a violin or a guitar string does. It's the same principle that some performers use to produce music by rubbing their finger on the rims of a range of tall glasses filled with water at various levels. The vibration makes each glass produce a different pitch because it is a different sized cavity. My mixer without any load gives me a very clean tone, almost dead on F#7 (about 2.96 kHz).

Probably the loudest sound in the kitchen comes from the air extractor device, especially older types or ill-fitted ones with tight curves in the extraction pipe leading to outside. They are generally fitted with a single or double horizontal fan, but most of the noise comes from the air rushing through it and through the filter. The fan has a larger diameter than the extraction pipe, so it compresses the air as it pushes it against the narrow pipe. This creates turbulence[1] just behind the fan, which makes a noise at a lower frequency, and this is exacerbated if the pipe has a tight curve just behind the fan.

A minor contribution is the high frequency noise made by air moving through a blocked extractor filter. The filter material is designed to be wetted only by oils and fats, but not by steam or water (its surface atoms are hydrophobic). If you tend to fry a lot, the filter will block fairly quickly and will emit a hissing sound as the air rushes past.

Fridges also make noise. This is due to the compressor at the back, which compresses the cooled gas as it returns from the evaporator inside. New fridges are very quiet, but since the compressor works thousands of hours every year, its internal bearings eventually wear out and it gradually becomes noisier.

The magnetron in the microwave oven makes an interesting noise when operating. When we switch on the oven, the cooling fan and the little motor rotating the glass plate start first. After 1–2 seconds you can hear the magnetron being energised as it makes a deeper noise. That's because the high voltage (nearly 2300 V) which sets up the electric field makes the cavity vibrate slightly.

Another major contribution to noise in the kitchen is actually the water tap. Because of water turbulence inside the pipe (and at its exit), all taps make some high frequency noise. On exiting, there is a laminar (smooth flow) to turbulence transition which adds to the noise, at low frequencies this time.

[1] Richard Feynman once said that turbulence is one of the big unanswered questions in physics. And Werner Heisenberg apparently once said that when he died, he'd ask God to explain quantum electrodynamics (or relativity) and turbulence to him. He was optimistic about getting an answer for the first, but not for the second. Physics and mechanical engineering students feel the same.

Sometimes, if there is a small amount of air trapped in the pipe, there can even be a loud bang in the pipe (at the highest point) on opening or closing the tap. This is the "water hammer" effect: because water is almost incompressible, a pressure release at the tap can send a pressure wave back up the pipe.

Dishwasher Power

The dishwasher has become an essential helper in the kitchen and, to its credit, probably washes dishes better than washing by hand. Its operation is relatively simple in concept. Water is heated, mixed with soap, and repeatedly sprayed at high speed onto the crockery, before being rinsed out by copious amounts of water. At the end, a high-surface tension water solution is sprayed on to reduce spots that may be left after drying.

The "soap" used in a dishwasher is actually a very caustic, sodium hydroxide-based alkaline cleaning compound which you don't want to get your hands wet with. It's extremely corrosive to all protein foods, but it also attacks glass.[2] In addition, dishwashers need some coarse common salt (NaCl) in order to soften the water before use, especially in hard-water regions. Softening entails replacing—by ion exchange—the metal ions in the water (Ca, Mg, and others) with sodium from the salt. This enables a better solution of the cleaning material and better wetting of the surface of the crockery and cutlery.

There are three moving parts in the dishwasher. The electric pump, which uses a paddle-type fan to push water to the top of the washer, and the two rotating wings with holes for distributing the water everywhere. The wings rotate by the water rushing out of the holes, in a nice demonstration of Newton's third law of motion,[3] and they produce a characteristic periodic whoosh, whoosh sound as they slowly rotate around. By the way, the water is heated by an embedded electrical resistance heating element, which accounts for most of the electricity needed to operate the dishwasher, or the clothes washer for that matter.

[2] Ordinary soda glass is chemically attacked (etched) by caustic soda (NaOH) so, after a few washes, ordinary glasses become foggy or mat. But even high quality "crystal glass" or even borosilicate glass is gradually attacked by NaOH. Nasty stuff.

[3] "Every action has an equal and opposite reaction"—used to accelerate rockets or planes through space by spewing combustion gases out the back of the rocket or the jet turbine of a plane. The rate of change of the momentum (mass × speed) of the waste gases is the force pushing the rocket or the plane.

Energy Waste—Tips and Tricks

Everything we do in the kitchen uses energy. Whether it is cooking or washing up afterwards, we may use some gas and certainly a lot of electrical energy. With the recent realisation that energy costs will go on increasing in order to enable a full scale transition to renewable electricity sources, it is worthwhile trying to find ways to minimise energy waste.

In our comparison between gas and electricity for cooking, we already identified some sources of energy waste. Gas hobs waste at least 40% of their energy if the flame is not properly controlled and limited to the underside of the pot or pan. In addition, if the crown is not spotlessly clean, combustion will not be optimal, which means the flame will be coloured and jump around, and also that its temperature will not be high enough. Moreover, it will produce some carbon monoxide and maybe other toxic gases alongside.

Electrical resistance hobs can waste substantial amounts of electricity if the pot or pan is not sitting properly on the hob due to warping. Metal hobs radiate a lot to the environment if the pot does not fully cover the hob, wasting more energy. Flat-topped glass-ceramic hobs also waste some energy because the heating coils sit under the glass–ceramic and radiate heat around them. Induction heating seems to be less prone to waste, as long as the pot sits properly and its size matches the size of the induction ring.

Another energy glutton is the simple kettle, if not used carefully. Because we use it so frequently during the day, even small savings here can translate into a major overall energy benefit. The most important mistake most people make is to boil a lot more water than will be used at that moment. Please don't look at the "minimum" water line. It's only there to remind you never to leave the kettle empty, and to enable the automatic shutdown mechanism (another bimetallic strip). For a cup of tea, just use a cup and a half of water, for 2 cups, 2 and a half, etc. Anything more is a waste of energy. Also, and this is important, don't wait for the auto shutdown to operate. Shut it down manually as soon as the noise has abated and the water is about to boil. Remember the best tea is made with water at 90 °C, not 100 °C.

The oven is probably the single biggest user of energy in the kitchen, probably more than the rest of the kitchen put together. The door seal, door thickness and quality, insulation quality, and size of oven all play a role in energy waste. Even with an extremely well insulated oven, the whole exercise suffers from the simple fact that the whole of the inside surface and air need to be heated up to the baking temperature before any real baking can commence in the cooking dish. Because food is not a good conductor of heat, it takes at least an hour for heat to penetrate into the heart of a piece of meat or vegetable. The rule of thumb is that it takes one hour for heat at 180 °C to penetrate to a depth of just 5 cm in cold meat (surface to core) in order to heat up the core to about 65 °C. So a normal 2 kg roll of meat with a maximum diameter of about 8 cm needs at least 2 h to cook properly.

Vegetables in the oven have another problem. Because of their high water content, most of the energy goes into heating the water before boiling point is reached, and this can take quite some time since water has such a high heat capacity. For many vegetables, 100 °C is not even sufficient for proper cooking in the oven. Potatoes, aubergines, zucchini (baby marrow), even tomatoes and peppers, all need higher temperatures (about 150 °C) to break down the external, fibrous skin and become fully eatable. And this can only happen when the water near the skin has evaporated, which needs a lot of additional energy to "feed" the latent heat of evaporation of water, as we saw before.

There are many tricks you can use to waste less energy when baking in the oven. First of all, you could try to heat up the food internally before putting it in the oven. The microwave oven is ideal for this purpose since, as we saw before, it heats up the inside of a piece of meat or a vegetable very efficiently and quickly. Just 6 min intermittent heating of a 2 kg roll of meat at about 2/3 power will warm it up to around 45 °C. This will then require about half the time to bake in the oven.

Stuffed vegetables can be baked in half the time if the stuffing is boiled for a few minutes before stuffing them. The vegetables themselves can also be lightly fried or boiled before filling them. Layered oven dishes like mousaka or au gratin benefit from initial light frying of the aubergines and potatoes and almost complete cooking of the meat filling before placing them in the oven. Vegan casseroles like briam should be baked after cutting all vegetables in small pieces and perhaps pre-heating them in the microwave oven.

Bread or cake baking can be speeded up if the dough or mixture is made dense by reducing the amount of water it contains. Dough should also be allowed to rise on its own at room temperature and covered before baking. Ideally, baking should start at a higher temperature (about 180 °C) and then reduced to about 150 °C as soon as rising is completed, which will save a lot of energy.[4] To be sure, small size bread-making machines are much less wasteful than baking bread in the oven, unless you bake 2–3 loaves together. By baking 2 or even 3 items together in the oven, you can save a lot of energy, although this is not always possible or desirable.

[4] Radiative and other heat losses increase exponentially with increasing temperature. Keeping an oven at 200 °C requires more than twice as much energy as keeping it at 150 °C.

Another major energy waste in the kitchen is use of an air-conditioner while cooking. If you want the amount of energy generated by cooking to be removed by the air-conditioning unit, then it will probably need to work continuously, negating any saving you may manage. It's probably much better to rely on the extractor fan and cool the kitchen down after finishing cooking.

An extractor fan will also use more energy if an excessive amount of sticky dust has collected in its bearings. This happens after years of cooking, especially if the filter is not changed frequently and there has been a lot of frying. After a few weeks of cooking, the filter will have adsorbed the maximum amount of oil it can hold, and after that, even if not blocked, any oily substance will go straight through the fan and into its motor. If you hear the fan struggling to get going and only gradually accelerating, it probably means that the bearings are clogged and it's time to oil it or change it.

Slow acceleration and fast deceleration (as though it is braking) of any motor is a sure sign of clogged bearings or sliding surfaces, a situation which uses up a lot more electricity. So, if any cooling room fan is showing these signs, it means that it is time to oil the bearings. Very thin, penetrating oil sprayed directly onto the fan spindle and allowed to soak and spread might work because it tends to dissolve thick oil.

Electrical power consumption

All electrical appliances and circuits consume energy and it is fairly simple to calculate the power in watts (energy per unit time) they use. All you need is to measure the current they draw and multiply by 220 V, since power is voltage times current. There are small devices that you can plug in to measure power automatically. They display current (in amperes) and power (in watts) drawn and energy consumption in kWhr, so you can calculate the cost.

When a device is under some stress due to friction, heat, or some obstacle, it needs (draws) more current to overcome the problem and operate correctly. Since the voltage remains the same, the energy needed to produce the same work is higher. You can see that by using the small device above connected to a fan. If you apply an obstruction to the fan you'll see the current (and power) shoot up. Don't do it too long though, as the excess current will overheat the wires and the motor.

Small energy savings can add up while cooking. When boiling water, e.g., for pasta, always start with the lid on the pot. Energy escapes with escaping steam (which happens at all temperatures, not just during boiling), so keeping the lid on will allow the water to boil more quickly as energy loses are minimized. Afterwards, for pasta, you have to keep the pot open (see below). The same does not apply for baking in a covered dish in the oven. A cover would delay the heating of the food inside the casserole dish since air is a very bad conductor of heat.

To save energy, use the kettle to boil water and then add it to the pot. The kettle element is embedded in its base so energy losses are minimal (good, well insulated, kettles have energy efficiencies exceeding 95%), so it's the fastest and most efficient way of boiling large amounts of water.

During cooking in the pot, keeping the lid on if possible also saves energy and you'll achieve the same results at a lower hob setting. This works fine with stews and rice but doesn't work when boiling pasta or legumes as the gluten and other binding compounds form a thick and strong froth barrier which tends to block the release of steam and lead to overflow of the froth past the lid and a mess on the stove.

When boiling anything in the pot, remember that it's the "time at temperature" that decides the completeness of cooking. If you can reduce the hob setting while keeping the food boiling nicely, you are probably just fine. The rule of thumb is to use a high setting until just before boiling commences, then reduce it so that gentle boiling continues until the food is half ready and then reduce it even further to a simmering setting till the end.

Energy can also be saved if you switch off the hob or the oven a few minutes before the expected completion of cooking. Both a normal electrical hob and the oven will retain a lot of heat after switching off, enough for cooking to be completed. This doesn't work well for flat-top hobs and induction-heating ones, which do not heat the hob surface.

The fridge in a hot kitchen can also be a major waste of energy. If the heat cannot be dissipated well away from the condensing coil at the back of the fridge, the compressor will work almost continuously. It will help if you have a fan in the kitchen mixing and moving the air around while the extractor removes as much as possible of the hot air. By the way, if you notice that the fridge compressor re-starts frequently or doesn't stop, accompanied by inadequate cooling inside, it may mean either that some of the coolant gas has leaked out or that the condenser is not dissipating the heat properly, perhaps because it's too close to the back wall and there is inadequate air circulation around it. This also happens if the fridge is fitted-in under cupboards and there isn't enough free air circulation above it.

In general, the biggest energy users in the kitchen are the heating elements wherever they are. The dishwasher is another major culprit. By rinsing out dishes immediately after use (even with cold water), you can use the dishwasher at a lower temperature setting, thereby saving a lot of electricity. The dissolving capability of the caustic material used in them means that even a temperature of 40 °C is sufficient for satisfactory washing of rinsed dishes.

Finally, energy (and time) savings are always possible if you can use the microwave oven instead of the oven or the hob. That's the main reason that so many ready meals are now offered for microwave heating. But a number of simple cooking jobs can also be carried out in a microwave. For example, I already mentioned cooking eggs in less than 1 min, but boiling sausages in their packaging (PP and HDPE are safe) in the microwave oven or even the kettle also saves a lot of energy. Potatoes cut to a thickness of up to 10 mm can be fully cooked in the microwave oven in less than 4–5 min. Remember though that MW energy is distributed—and absorbed—in all the food that can absorb it (mainly water). So the larger the amount of watery food you place in the MW oven,[5] the longer it'll take to cook it. There is almost a linear relationship between the amount of water and the time. For example a small cup of soup (150 ml) from the fridge will warm up well in about 1 min but a large plate of soup (300 ml) will need 2 min to warm up well. In the case of soup though, I should comment that, for larger amounts, an induction heating hob may be quicker and cheaper than even a MW oven.

Hammer and Spark

I'm sure almost everyone has used a spark lighter at some point in the kitchen or elsewhere to light a gas flame. Some are hand-held, others are built-in next to each hob. They exploit a very interesting physics phenomenon called "piezoelectricity" (Greek for "compression electricity"). When you compress strongly but momentarily a piezoelectric material—striking it with a small hammer is best—produces a large potential difference between one side and the other, and this can be used to produce a sharp spark. Such lighters can generate more than 20,000 V which is more than enough to produce a good spark to light a gas–air mixture.[6] Nowadays, many modern gas ranges use electronic lighters instead of piezoelectric ones. Here, a high voltage arc (spark) is generated using a "Tesla coil" or other system in which a small voltage through a few windings in a transformer induces a very high voltage in an adjacent coil with thousands of windings. This creates a spark a few mm in length with any metallic part of the hob (the "earth"), as it causes a dielectric breakdown of the air. The same principle exactly is used in those mosquito-killing rackets - I love them.

---

[5] Most MW ovens output about 800 W of MW power with a total power efficiency of around 60–70%.

[6] The amount of current (number of electrons) that jumps across is tiny: a few μAmps at best, so there is no danger if you touch it accidentally.

Natural piezoelectric materials are quartz and topaz crystals, but most good piezoelectric devices use polycrystalline lead zirconate titanate (PZT, an oxide ceramic of lead with zirconium and titanium), which is very sensitive and stable.[7]

Electrical transformers

Transformers convert one voltage to another by electromagnetic induction. The supply voltage is connected to the "primary" coil and the current flowing produces a magnetic field. This is then sensed by the "secondary" coil, in which a smaller or larger) current is induced, depending on the ratio of the number of turns between the two coils.

Transformers are found in all electronic systems and many electrical ones, as circuits usually work with smaller or larger voltages than the 220 V supply. They are also used to isolate one part of the circuit from another for safety and to reduce interference. For example, garden lights, because of the high probability of short-circuits due to rain, are always supplied via a transformer where the primary and the secondary have an equal number of turns. In this case, the secondary voltage is also 220 V and the lights are isolated from the supply.

Piezoelectric materials are used in many other applications in the kitchen too. The electronic clock (or watch), still popularly called a "quartz clock," since piezoelectric quartz crystals are still used in them, is a ubiquitous application. Good quality clocks use PZT (or other piezoelectric crystals) in the inverse mode: a small voltage applied across its ends produces a small momentary nudge which moves a tiny cog wheel to advance the second hand, one second at a time.[8] It's not surprising that after a few million such nudges, the crystal cracks and the clock stops. PZT is tougher and lasts much longer than quartz for this reason. By the way, the buzzer of your oven clock also works with a large, flat piezoelectric polycrystalline disk. A repeating electric signal gets it to vibrate loudly (and irritatingly) at the set time. The old alarm clock next to your bed did the same.

[7] Lead on its own is known to be toxic, but here it is part of the compound and, in any case, such lighters don't come in contact with any food.

[8] Timing is provided by another crystal which operates (quantum mechanically) at a particular frequency and gives a voltage signal every second.

Searching for fish (or submarines)

A normal military or police radar works with microwaves (electromagnetic radiation) which travel at the speed of light. But this cannot work in the sea, so ultrasonic radars are used instead.

To detect a submarine or a school of fish, the transceiver sends a "beam" of sound and waits for the reply. Knowing the speed of sound in water (about 1500 m/sec), it can calculate the distance and, by scanning, it draws up a 3D map.

The observing vessel must be stationary or move slowly to get an accurate picture, especially if the object is moving. The speed of sound is also affected by colder water and water pressure, both of which increase the speed.

For comparison, the speed of sound in air is about 330 m/sec, and through concrete it is about 4000 m/sec.

If you use your telephone in the kitchen (who doesn't?), then you use another two applications of piezoelectricity. Modern phone microphones and speakers (and earphones) use the piezoelectric effect. Speech vibrates the piezoelectric crystal in the microphone, which then sends corresponding electric signals to the audio amplifier in the phone. When a signal is received, the reverse effect translates it into sound in the piezoelectric speakers.

Finally, although not necessarily in the kitchen, some burglar alarms use small ultrasound radars based on the same principle. A beam of ultrasonic pulses is generated by a piezoelectric crystal vibrating at high frequency (usually 40 kHz), which bounce off the opposite wall and come back to a similar crystal where a little computer computes the time taken and thus the distance to the wall. When an obstacle obscures the beam, the computer calculates a different distance and raises the alarm. Exactly the same piezoelectric "ultrasonic radar" principle is used by boats and submarines to map out the rocks in the sea below them and monitor the movements of other submarines in the sea. Fishing boats and research ships use them to locate schools of fish and the odd whale too.

Weighing up

Modern kitchen scales operate on the basis of either one of two main principles. The first, and most frequently encountered, uses a metallic cantilever beam[9] which deforms slightly when a load is placed on the scale. The deformation increases the electrical resistance of a very sensitive copper resistor wire

[9] A small bar of metal that is bent by loading the scale.

glued to the beam, and the resulting change in the voltage applied across it is then registered. By calibrating the resistance change with a known load, the load placed on the scale is determined. This is the most widespread method and the same principle is utilised for measuring loads in industry, at home, and in the science lab.[10] Loads up to many tons can be determined, for example in truck weighbridges.

The second type of scale is more advanced and more sensitive and operates by the piezoelectric principle. A load placed on the scale deforms the piezoelectric crystal (or, more usually, a polycrystalline ring), which produces a voltage signal proportional to the deformation. By calibrating the voltage change with a known load, the load placed on the scale can be determined.

Interestingly, variations in the ambient temperature can affect both measurements, but only to a very small extent. The precision in weighing required in the kitchen, probably around 1 g, is too large to worry us, so we don't have to take that into account. For comparison, chemical lab scales can measure with a precision of 0.0001 g and specialist scales exist with a precision 100 times smaller so they take into account temperature variations..

Tic–Toc

In addition to electronic clocks, wind-up timers are still used in many kitchens. I do, as they do not need a battery. In electronic clocks, we need a source of energy which will regularly help to push the second hand lever forward. In a wind-up clock, that source is the energy stored in a hard metal spring. When we wind up such a clock (but also a grandfather clock or a child's wind-up toy), we are putting a coiled metal strip under a load (mechanical tension). The more we wind it the greater the load. The total energy stored[11] is (roughly) the maximum load times the amount of deformation of the spring. In order to use this energy, we allow the coiled spring to unwind gradually (using a toothed ratchet and gear mechanism), adjusted so that it pushes the second hand lever once every second. This turns the second hand, which is meshed with the minute hand and then to the hour hand, if there is one.

Since we are discussing time, it's worth remembering that time is always of the essence in cooking. But using the durations mentioned in cooking recipes can easily lead to culinary disasters. Every oven and even most hobs

[10] Bathroom scales operate using four of these cantilever beams, one in each corner.

[11] The original energy source is our muscles from the metabolic conversion of food + oxygen, before that the growing plant, etc., all the way to the Sun.

are different from each other. One should use recommended times and temperature settings in recipes with a pinch (or more) of salt.

Ovens are especially prone to variations in cooking times. The internal size of the oven, the power output of the resistive elements, and the insulation quality all play a role. The seal on the oven door may be ill-fitting so the time taken for a casserole or a roast may be a lot longer (wasting energy in the process, of course). The use of the air-distribution fan inside the oven may help in distributing the heat more evenly, but it can also expel hot air, taking energy with it.

New cooks are particularly prone to timing problems, based on recipe suggestions. My simple answer to such worries is to trust your eyes: the only way to be sure that cooking is proceeding well is to observe the result and make adjustments as necessary. Everyone learns cooking by making mistakes. And sometimes "mistakes" turn up to be delicious. Serendipity is a cook's best friend.

Kitchen on Fire

Dangers abound in a kitchen. We've already seen what can happen when we overheat oil in a fryer and throw many potatoes in at the same time. The sudden creation and emergence of thousands of bubbles will swell the oil and can very easily cause it to spill over the pot, whereupon it will catch fire immediately, either in the gas flame or on the hot hob.

Cool oil spilt over a hot hob or even in a flame will not easily catch fire. But hot oil already contains lots of energy and it needs only a little bit more to evaporate, whereupon the vapour can react with atmospheric oxygen and catch fire.

Once a fire has started the best or only way to put it out is by restricting its access to oxygen, i.e., air. Once oxygen is eliminated from the reaction equation, the fire will automatically fizzle out. Every kitchen should have a fire blanket handy that one can throw on the fire in the event of an emergency. Failing that, a wet towel is the second best choice. Never beat or throw water on an oil fire as all you'll do is to spread it around.

A major source of danger in the kitchen is the gas itself, and in particular gas hobs that work with bottled LPG gas. These systems do not have a fail-safe system for shutting down the gas flow so they should not be left unattended. If the flame goes out and the gas leaks into the room there is a major risk of toxicity or even an explosion.

This is not a problem with most modern installed ranges since they have flame-out safety systems in each hob which shut down the gas flow if the flame goes out. They work by detecting the temperature at the crown using a thermocouple or other temperature sensor (e.g., a bimetallic strip working electrically or mechanically) and de-activate an electromagnetic or mechanical valve which is keeping the gas flow pipe open. This is a fail-safe system as the valve is spring loaded and normally closed, meaning that, if the flame goes out and the device fails, the flow stops.

Other sources of fire in a kitchen (and elsewhere) are electricity power outlets. Overloaded plugs with too many appliances attached may be forced to supply a lot of current at the same time and will at some point overheat, possibly melting the socket or the plugs and causing a fire. Nearly all normal home outlets can supply a maximum of about 12 amps total current, which means about 2640 W of power at 220 V AC,[12] which is about the power rating (power drain) of just one large room heater or a large electric hob working at maximum. To check whether a socket is overloaded one can add the power ratings of all appliances (always indicated on a label on the appliance) connected to the same socket to ascertain whether the plug can provide the total current (or power) needed by all of them.

But even if the total power consumption is not above the total rating of all appliances connected to one socket, a fire danger still exists for ill-fitting plugs in the sockets. It is not unusual for plug pins to make weak contact with the contacts in the socket leading to localised arcing at the point of bad contact. This arcing generates very high localised heating (which may or may not be audible as a faint high frequency hiss) and will eventually overheat both the plug and the socket. The older type of outlets and plugs were made of hard bakelite (a thermoset polymer based on phenolic) which was resistant to high temperatures (above 200 °C) and gave off a distinctive "fishy" smell when overheated, thus warning of imminent danger. Modern outlets and plugs are unfortunately made of less durable plastics (ABS or similar), which melt or carburise without much warning. To avoid potential fires, always make sure that all plugs make good contact in the sockets and replace any that appear melted or damaged.

[12] In countries with 110 V mains, sockets can supply up to 24 A (80% of 30 A), which comes to the same maximum power rating (2640 W).

Indelible Stains and Miracle Workers

When I discussed the operation of the dishwasher, I mentioned that the cleaning agent used there contains a caustic material, probably based on sodium hydroxide or something equivalent. But a washing-up liquid (and all soaps) could hardly contain such caustic compounds since it would wreak havoc on our hands when we do the washing-up.[13] What it does contain are substances that, on the one hand, increase "wetting" (adhesion) between water and oil (a type of emulsification), and on the other, reduce the surface tension.

The overall mechanism for cleaning with a soap does not rely on dissolving the dirt but on dislodging it from the surface and is quite interesting. The emulsifiers in ordinary soaps are sodium—or potassium-based "fatty acids" (produced from oils by "saponification"), so they are partly organic and partly inorganic molecules. This means they have both a hydrophobic and a hydrophilic side, so are strongly polar. Because of this they join up with water and at the same time break up non-polar oil molecules. If a stain is wet, it is removed easily. If it is dry, the water side diffuses into and around it and gradually "lifts" it off the surface, so that it can be removed. However, if the water is acidic, this mechanism cannot work as it neutralises the emulsifier. Modern washing-up liquids (and some liquid soaps) contain other substances that are able to work in both acidic and alkaline environments. Froth helps both reactions. It mixes in oxygen, which oxidises the stain and weakens the bonds.

Cleaning agents used for removing polymerised and cross-linked stains on the outside of frying pans and on the inside walls of ovens contain all these as well as caustic compounds which, at high enough temperatures, dissolve the stains and then dislodge them as above. With a lot of elbow grease too.

[13] Old washing-up liquids did contain such skin-eroding substances. Thankfully no more.

Why equal mass?

The first of Newton's laws of motion defines the inertial mass of the body as the property which resists change in speed and is the constant of proportionality between acceleration and force in Newton's second law. Interestingly, it happens to have the same value as the mass that is defined by Newton's law of universal gravitation, a fact which has been confirmed over and over again to an accuracy of better than one part per trillion. It simply means that it's impossible to tell the difference if you are in a lift falling freely or being accelerated upwards with the same acceleration. This is known as the "weak equivalence principle" and is a major foundation of physics.

Oddly enough, we don't actually *know* why they are the same and we just accept it as a fact. If you find the reason why inertial mass and gravitational mass are the same, you'll probably become extremely famous and get the Nobel Prize. Einstein started from this weak equivalence principle to develop his famous general theory of relativity, which explains gravitation.

Sucking Things up, and Other Bits and Pieces

Looking around the kitchen, I can see a number of other smaller and larger devices and appliances that use physics principles to operate. The vacuum cleaner simply creates suction by reducing the pressure, creating a partial vacuum in front of the fan (or a turbine, just like jet planes), whence surrounding dusty air rushes to fill it and is then forced into the fan blades and from there into the filter bag. This holds on to the dust and other particles and allows the "cleaner" air to escape.[14] Dyson-type cleaners do not use a filter bag. The air that is sucked in is made to revolve around a cylinder, which forces the particles to collect in the centre of the cylinder by the same principle that makes tea leaves collect in the centre of the cup (see "Storm in a tea cup" earlier).

A blender is simply a system of sharp rotating blades chopping through food at high speed. It's the speed that does all the work and the inertia of the food that allows it to be cut so easily, even with blunt blades. Inertia is a fundamental property of bodies and is proportional to the mass of the body. It's the reason your car refuses to budge and needs a big push to get it to start moving, but is then relatively easy to keep going. In space, it is the reason why the International Space Station moves just a little every time an astronaut pushes against it.

[14] Except the very fine particles (most of them nano-sized) which produce that characteristic smell, irrespective of what we have sucked up, by stimulating our olfactory sensors at the nano-level.

The same principle works in any other motorised chopping device, for chopping pepper corns or anything else. For such small items, the speed has to be much higher since their inertia is so much smaller.

I can also see a small digital thermometer on the wall showing the temperature inside and outside the kitchen. Both sensors are resistive and we discussed them earlier. The small screen works by segmented "liquid crystals" which change light absorption characteristics every time a small voltage is applied on them, controlled by a tiny dedicated electronic chip inside.

Finally, a sponge or ultra-water absorbent clothes are made with extremely hydrophilic materials with millions of very fine capillaries. The combination of the surface tension of water in capillaries together with the hydrophilic nature of the material makes it able to absorb many times its own weight in water.

Post Script

And so on and so forth. I could continue with the physics of the materials the walls are made, the turbulence of the tap water, the movement of air, and why that metal rack gets sticky so quickly,[15] along with so many other phenomena, but I have to put a stop somewhere. Physics is indeed everywhere in the kitchen.

That's it. I have been looking around the kitchen and scribbling notes on small pieces of paper about various phenomena I have observed and thought of for over two years now. I do hope I have included and discussed all those observations clearly enough and hence that avalanche of little notes can now be recycled.

I have enjoyed writing this book tremendously. I hope you have enjoyed reading it too. All that is left is for me to thank you very much for reading this far and to wish you many happy hours cooking and observing! And of course, never stop being curious about nature!

[15] It's because of partly polymerised oil from boiling pots settling and condensing on cold surfaces and gradually, over time, cross-linking.

Further Reading

There exist a good number of popular science writings dealing with many of the phenomena we discussed in this book, both online and in print.

Some very well written books I have enjoyed on these and related subjects, and I hope you will too, include (in no specific order):

1. "Storm in a Teacup: the Physics of Everyday Life" by Helen Czerski, 2017
2. "What Is Life?" by one of the giants of QM, Erwin Schrödinger, 1944, with Mind and Matter and Autobiographical Sketches and a foreword by another great (contemporary) thinker, Roger Penrose
3. "The Greatest Show on Earth: The Evidence for Evolution" by Richard Dawkins, 2009
4. "Into the Cool: Energy Flow, Thermodynamics, and Life" by Eric D. Schneider, 2005
5. "The Vital Question: Energy, Evolution, and the Origins of Complex Life" by Nick Lane, 2015
6. "Helgoland: Making Sense of the Quantum Revolution" by Carlo Rovelli, 2020
7. "Six Easy Pieces", essentials of physics by its most brilliant teacher, Richard P. Feynman, 1994
8. "Chemistry at Home: Exploring the Ingredients in Everyday Products" by John Emsley, 2015

Regarding online popular physics and chemistry writings and news, I frequently visit and enjoy the following sites:

1. https://www.iop.org/
2. https://phys.org
3. https://rsc.org
4. https://www.newscientist.com/
5. https://www.edge.org/
6. https://www.scientificamerican.com/
7. https://www.the-scientist.com/

but there are many others.